Streamlining & Car Aerodynamics

By Jan P. Norbye

MODERN SPORTS CAR SERIES

TAB BOOKS

Blue Ridge Summit, Pa. 17214

FIRST EDITION

FIRST PRINTING—NOVEMBER 1977

Printed in the United States
of America

Library of Congress Cataloging in Publication Data

Norbye, Jan T.
Streamlining — car aerodynamics.

Includes index.
1. Automobiles—Aerodynamics. I. Title.
TL245.N67 629.2'31 77-18191
ISBN 0-8306-9997-X
ISBN 0-8306-2037-0 pbk.

about the author

JAN NORBYE is one of the most respected automotive journalists in the world. Born in Norway, for many years a resident of the United States, and now living in France, Norbye is international editor of Automotive News. The title was created for Norbye and he has transformed the international coverage of this weekly bible of the industry. For four years, Norbye was technical editor of Car and Driver magazine in New York. From the mid-sixties to the mid-seventies, he was automotive editor of Popular Science, and is largely responsible for most the incisive and helpful road tests in publication. He is the author of *The New Fiat Guide* and *Chassis Tuning* in the Sports Car Press series and has written *The Wankel Engine* and *The Gas Turbine Engine* published by Chilton. With his thorough technical grounding and a capability in many languages, Norbye is a notably successful popular journalist, seemingly able to tame any complex subject. He washes the words over you in an easy, comprehensible and entertaining sytle, making you confident you have learned a lot and learned it from the best. Norbye is a member of the Society of Automotive Historians and an affiliate member of the Society of Automotive Engineers.

contents

introduction

What is streamlining? It became a catchword for fashionable design in many areas back in the 1930s, long before automotive aerodynamics was a well-understood science. Industrial designers claim to have streamlined office machines, vacuum cleaners, and bathroom fixtures.

In modern language, streamlining has come to mean other things. You "streamline" an organization, for instance. That means cutting out loss-producing functions and laying off non-productive personnel. By that definition, it is fairly analogous to the principles of streamlining a car.

A "streamline," according to the *Encyclopedia Britannica,* is an imaginary line that indicates the direction of fluid velocity at each point along its path. *Dictionary of Science* says that it is a h pothetical line through the air that is tangential to the flow at all points.

If that seems confusing at first, it will all become clear when we start looking at how the air flows around a moving car; that a car's motion disturbs the air, and streamlining consists in shaping car bodies to minimize this disturbance.

Of course, you have a certain idea of what a streamlined car is. You know that the 1934 Ford V8 was not, while the 1934 Chrysler Airflow was. The difference lies in the way the air around them behaves. The fact that air flow isn't something you can see with the naked eye as it occurs under real-life

Citroen SM set something of a new standard in the streamlining of practical four-seater grand touring cars. Air intake is below the bumper, and headlights are enclosed in Plexiglass covers.

Vehicle Road Load Horsepower Requirements

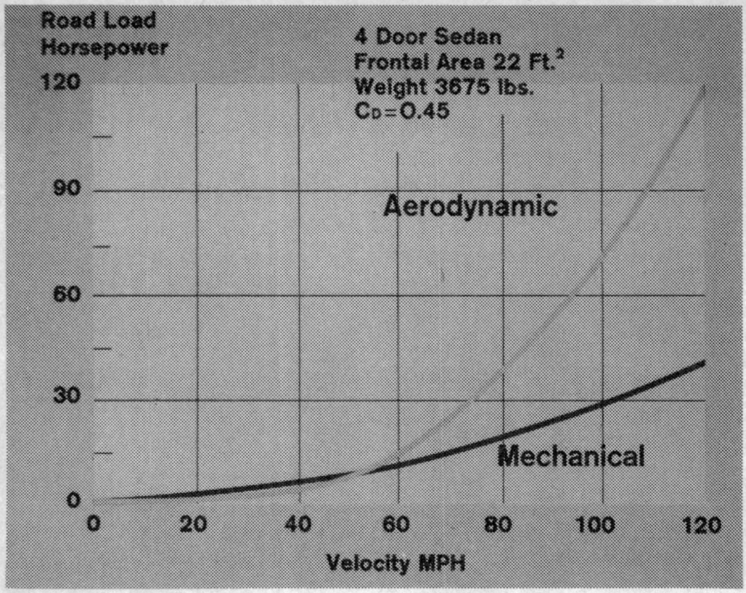

This graph shows how aerodynamic drag becomes the main force opposing the car's travel at speeds of about 50 mph.

conditions, has been the biggest problem over the years. Ways to visualize air flow have been developed, however, and we will examine test methods and measurements in a later chapter.

Nature has created remarkably efficient streamlining. Just look at a seagull or a porpoise, and you know you are seeing creatures designed to move fast and without waste of energy.

Streamlining Reduces Fuel Comsumption

Driving a car at high speed with low fuel consumption is everybody's goal today. That's why the same kind of streamlining that works so well for seagulls and porpoises is so important for car design. In technical terms, streamlining can be defined as *reducing aerodynamic drag*. Don't let those words stop you. Aero means air and dynamic means motion. Aerodynamics, therefore, is the behavior of air in motion relative to a body. Drag is self-explanatory. For the car, it's a force that tries to hold it back when it wants to move. Drag resists speed. To overcome drag, more horsepower must be applied, and that means consuming more fuel.

With lower drag, the car can go farther on a gallon of gas at a given speed. By developing a body with lower drag, the designers provide the engineers with the possibility of using a smaller and more economical engine without giving up performance and speed.

Various forces act together to keep the car from moving. Air resistance is only one of these forces. There are also mechanical forces, which are better described as "losses." You lose some power in the transmission, due to gear and bearing friction. You lose some more due to the friction between the tires and the roadway. This is called the rolling resistance, which depends first and formost on the load the tires support, and secondly on speed.

At low speeds, mechanical losses are greater than aerodynamic losses, or air drag. As speeds increase, the air drag becomes the more important. While mechanical losses increase proportionally with speed, air drag is a function of the square of the speeds.

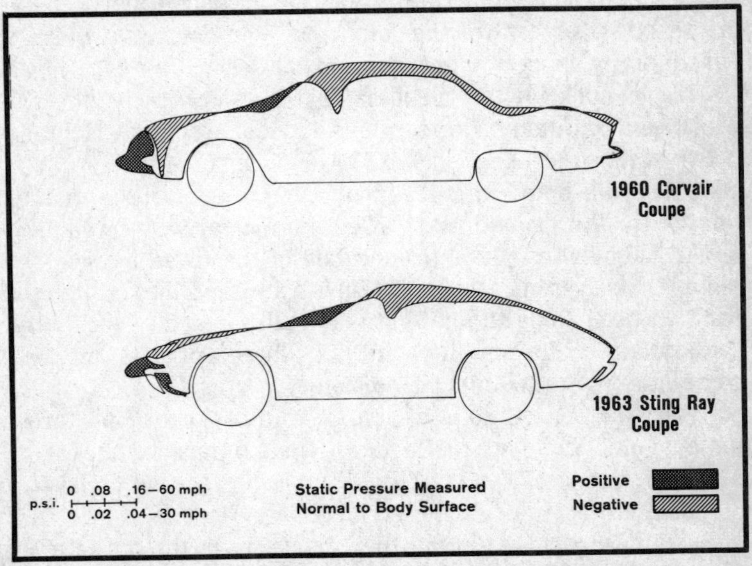

Pressure distribution on two different cars show both similarities and contrasts. Corvair nose builds up high pressure (= higher drag) while Corvette nose splits the wind with minimal drag. Corvair rear deck lid has small high-pressure area where louvers are intended to let cooling air escape—a bad error.

Air drag increases in a geometrical progression. On the graph, you can see the curve rising at a steeper and steeper angle with higher speed. For each car there comes a point when a 10-mph speed increase will double the air drag. At normal cruising speeds, air drag is the most important single influence on a car's performance and fuel economy.

Limited Help from Airplane Design

When the first streamlined cars were designed, car aerodynamics was a new and not fully understood science. At first, a streamlined car was thought of as borrowing aircraft techniques in its body design. Elements of aircraft shapes were frequently incorporated in car bodies, sometimes with awkward results, and sometimes to great benefit. Later, we learned that principles of aerodynamic design for cars are completely different from those that apply in aviation. This is largely because the car is "grounded." It runs on wheels, with its body always in close proximity to the ground. As an aerodynamicist would say it: The car moves on the interface between a solid and a fluid, while the airplane, when flying, is entirely surrounded by a fluid. The ground effect will be discussed in several of the following chapters, for it plays a part in many aspects of air flow and car design.

If you look on the airplane fuselage as a body, the car becomes a half-body. It resembles a fuselage split horizontally in two at its widest section, with wheels put under it.

This half-body has freedom to move in three planes relative to the ground plane. The motions are yaw, roll, and pitch. Each takes place around one of the three axes in the motion axis system, and each axis is a straight line through the car center—longitudinally, laterally, and vertically. Aerodynamic forces act in all three planes, in any combination, according to conditions.

Air flow is of course not two-dimensional, but three-dimensional. The body deflects air around its sides as well as over the top and below its belly.

As soon as the car begins to move, there is an air flow around the body. Assuming there is no wind, the direction of this air flow is directly opposite to the car's travel path. Head-on, that is. In real-life conditions, there is nearly always a wind, however. And it is rare for the wind direction to coincide with the car's travel path. It is usually a sidewind,

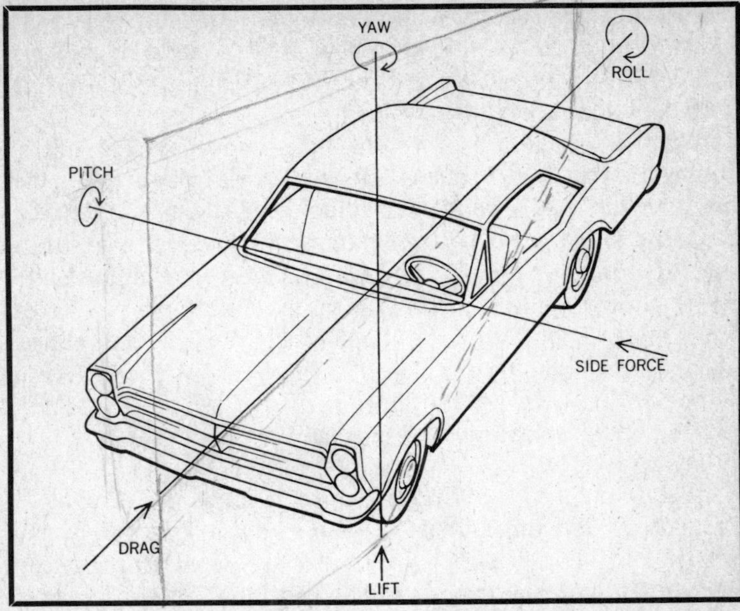

Aerodynamic force diagram shows the three axes along which winds can act on a body, and the three planes in which the car has freedom to move: roll, yaw and pitch.

varying from a few degrees of yaw to blowing right across the beam. The effects of sidewinds are a vital part of car aerodynamics. Streamlining a car means reducing the effects of both headwinds and sidewinds.

Before we go further, it's important to realize that the car also depends on air flow for vital functions: engine breathing and cooling, brake cooling, and interior ventilation. There's an *internal* airflow as well as the wind all around the car. What happens to the air inside the car, and how it escapes, can have considerable effect on the overall drag. What's more, the external airflow comes into play in many areas other than fuel economy and speed. Winds affect windshield wiper operation, exhaust gas dispersion, dirt accumulation on the body, and wind noise inside the car. Last but not least, the body shape and the air flow around it also influence the car's stability at speed.

40 Years of Stagnation

In the first 40 years that elapsed after Chrysler launched its Airflow model, no measureable air drag reduction was achieved in the average American car, despite the vast expansion of aerodynamic knowhow. Now, under the pressure of government-imposed fuel-economy standards, the Detroit car is becoming more streamlined.

Why did it take so long? Mainly because, for mass-produced cars, cost considerations had veto power over the best technical and scientific arguments of engineers, stylists, and aerodynamicists. An aerodynamically bad idea that saved a nickel a car was something that management would rush to adopt, while a genuine improvement in streamlining that cost a dime per car would be kept out of production unless management could be convinced it would sell more cars. Even on the part of the best-intentioned people in the industry, there was a certain resistance to accept the plain facts of car aerodynamics.

As late as 1949, P. Weychan wrote in *The Motor* (London): "Outside of certain obvious principles one might safely say that, at present, designers tend to look past this feature [aerodynamics], because the ratio of the time when a car is driven at speeds of wind-resistance importance, to the time of that car's moving life, would not justify the additional research and production expense."

And in 1953, when Edward N. Cole was chief engineer of Cadillac, he was quoted in *Cars* magazine as saying: "Cadillac tests show that until an automobile has attained a speed of more than 60 mph there is nothing to be gained, from the standpoint of performance or economy, in the aerodynamic characteristics of the car. Consequently, as the average speed of majority of passenger cars, throughout the car's life, is only 28 mph, you might as well be riding around in a car built like a hay wagon."

Just what are the facts? I'll explain the various drag factors and their importance in the next chapter. As for the argument that cars don't travel at high speed for any significant portion of their lieftime, I can only say that the car's average speed throughout its life is a totally meaningless abstraction. What counts is the energy consumed in moving the car and you only have to consider the millions of cars running at high speed on our interstate highways and other modern roads at any moment of day or night to know that streamlining can save vast quantities of gasoline.

chapter 1
about air and drag

You may be wondering why the air produces a drag on the moving car. Could not the air instead assist it on its way, or at least, let it through toll-free? No. You have just run right smack into the laws of nature, or more explicitly, one aspect of the laws governing air flow that is known as Bernoulli's theorem.

Daniel Bernoulli was a Swiss mathematician who published a major work on the behavior of fluids, "Hydrodynamica," in 1738. Bernoulli's theorem says that the sum of the kinetic energy in an airstream and its pressure is constant. What that means in practice is that a drop in pressure will result if the air flow is speeded up.

This not only explains drag, but is also the basis for all aerodynamic improvements in moving bodies—ships, aircraft, and automobiles.

To fully understand it, you will need first to make up your mind what air is. Air is a fluid, and as all fluids, it has a viscosity. That's not just something you get in motor oil. It's a measure of a fluid's resistance to relative motion within itself. This leads you to the conclusion that the air has a certain internal friction.

Air Gets Down to Push and Shove

Basically, air is a collection of molecules moving at very high speed in what you can regard as empty space. Not all

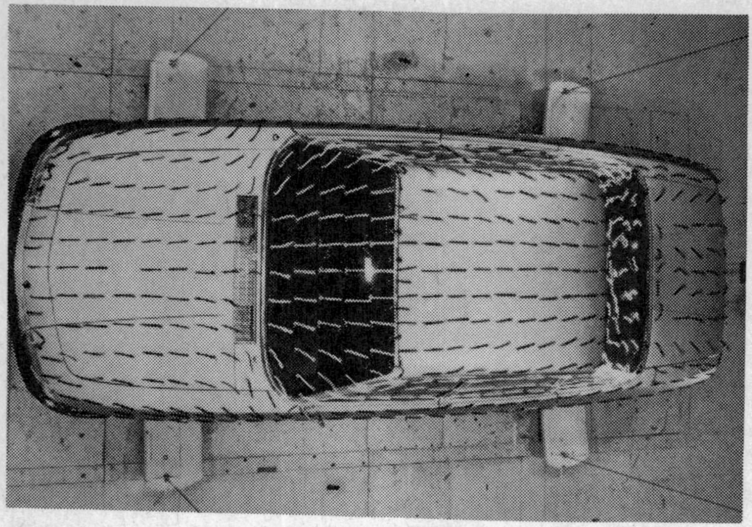

Remarkably good streamlining for a notchback sedan was demonstrated in this wind-tunnel test of the 1967 NSU Ro-80. The front end is particularly successful, but some turbulence occurs across the rear window. Its C_d is 0.355.

molecules are the same size or travel at the same speed. Oxygen molecules move at a mean speed of about 1,100 mph. Hydrogen molecules move about four times faster.

How does this affect the car? Well, the car is subject to pressure from the air. Static air pressure acts on a stationary body, and static air pressure against a body can be defined as the cumulative effect of countless molecular impacts. Dynamic air pressure acts on bodies in motion—or, in other words, it is what you get when there is relative air flow over the body. It is important to recognize the expression *relative air flow*, for it dismisses the question of whether it's the car or the air that's moving. In the study of aerodynamics, it does not matter. What counts is the way the air and the body come together and separate.

Dynamic air pressure differs from the static only in the sense that instead of random impacts you are dealing with impacts directed by the energy of a moving column of air.

In the final analysis, it becomes a matter of organization. The kinetic energy of air molecules is organized in one of two states: random or directed. Now we know from Bernoulli that faster air flow will reduce the pressure, which is probably just the opposite of what you would expect.

The Culprit Is Negative Pressure

At this point it is necessary to bring in the idea of negative pressure. You are probably familiar with the concept as it works in carburetors (manifold vacuum is a form of depression, or negative pressure). But negative pressure does not have to be enclosed—it can also exist in open air. And negative pressure can and does act on moving bodies such as cars. Drag is mostly negative pressure.

A picture of the pressure distribution on a moving car will show which areas are rammed by the air flow, where maximum positive pressure is generated, and from which areas the air is sucked away, where negative pressure values are highest.

As we know, from Bernoulli, the pressure is lowest where the air speed is highest. The acceleration in air speed is caused by the moving body itself. What happens on the exterior surface of the car is easily explained.

Quite simply, the air speeds up as it goes around an obstruction, doing all it can to keep up with the main mass of the air which passed by, at some distance, in a straight line. The deflected and accelerated air exerts a drag on the entire surface swept by it. The word drag implies holding back from behind rather than blocking from the front. And that gives you the correct impression. The ram effect of the airflow being displaced by the moving car is relatively low. Thanks to Bernoulli's theorem we know that the ram pressure we can obtain from an airstream at 50 mph is only about 0.05 pounds per square inch. Even at 100 mph it is not a major force—only about 0.15 pounds per square inch. It is the negative pressure that produces the major drag—acting on the roofline, deck, rear window, and body sides. These are the critical areas, where streamlining is most needed.

Streamlining Overcomes Drag with Less Power

How much power is required to overcome air drag at any given speed can be calculated when you know the frontal area and drag coefficient of the car. Frontal area is defined as the silhouette that meets the airflow head-on. It's a two-dimensional concept: the height-by-width surface that faces a horizontal air flow. The coefficient of aerodynamic drag is a symbol of the body shape without regard for its size. It is a meaningless figure in itself, but a valuable basis for comparison between different shapes.

Camaro was GM's first body developed in the wind tunnel. This design has turbulence around windshield pillars and wheel cutouts. Enormous radiator air intake makes for high internal drag.

A vertical plate facing an airstream has a drag coefficient of 1.17. An ideal shape, suspended in mid-air, has a drag coefficient of 0.04. Those are the extremes we have to deal with in discussing automobile streamlining.

At low speeds, frontal area is responsible for half of the total air drag. At higher speeds, it is the drag coefficient that dominates.

The formula for calculating the total drag of a car is easily understood even if you are not mathematically inclined.

$$D = C_D \times F \times A \times V^2$$

where

D = aerodynamic drag (in pounds)
C_D = drag coefficient
F = frontal area (in square feet)
A = mass density of the air flow $\left(\frac{\# \sec^2}{ft^4}\right)$
V = vehicle speed in mph

An example will show how this works. Take the 1934 Ford V8, which had a frontal area of 22.5 square feet and a drag coefficient of 0.60. Assume normal air density, and take a speed of 55 mph. This gives:

$$D = 0.60 \times 22.5 \times 0.0025 \times 55^2$$

with the answer D = 102.1 pounds.

For comparison, take the 1956 Citroen DS-19, which was one of the best examples of its period, with a drag coefficient of .311 and 21 square feet of frontal area. The calculation is

$$D = 0.311 \times 21 \times 0.0025 \times 55^2$$

with the answer D = 49.4 pounds.

With nearly the same frontal area, the more modern car had less than half the drag at the modest speed of 55 mph. In terms of power consumption, the Citroen needed only 13.2 hp to overcome air drag while with the Ford, it took 27.2 hp.

Every Car Wears an Invisible Coat

Total air drag can be broken down into five separate components, but before we can examine the various drag factors, it is necessary to discuss a phenomenon called the *boundary layer*.

Two views of the British Leyland Zanda. Its wedge-shape concept was developed into the TR-7, but the Zanda was a midships-engined design, while the TR-7 has a front-mounted engine.

The boundary layer is a thin layer of air that is not part of the flow, but adheres to the body. It is a viscosity phenomenon. Air particles that come into contact with the solid surface of the car body tend to adhere to it, and this adhesion may be stronger than their cohesiveness with other air particles. The boundary layer is not one uniform sheet of air, but rather an air coating of variable thickness around the car body.

Between the particles in direct contact with the body surface, and the outside airflow, are several layers in which particles are continuously changing places, from an inner layer to an outer, and vice versa. As long as there is an orderly, laminar flow in this coating, the surrounding air is not greatly disturbed, and air drag on the body will be low.

The boundary layer is semi-isolated from outside forces, with little momentum of its own. However, it is prey to changes in the pressure distribution outside it, and can easily be separated from the body surface. When separation occurs, you no longer have streamlining but turbulence.

The boundary layer gets its start at the front of the car, at the stagnation line. The stagnation line is defined as the line where the air that has to pass under the car separates from the air that will pass over it. At first, the boundary layer is very thin. Thickness gradually builds up, due to viscosity in the air.

When the Boundary Layer Is Broken

At some point, there is a transition from laminar flow in the boundary layer to turbulent flow. The farther from the front of the car this occurs, the lower the drag on the body, for that means larger areas with low-drag laminar flow, and smaller areas with turbulence. The separation is inevitable; only the exact point at which it occurs varies in accordance with body shape, speed, and air density.

But not all boundary separations are equally turbulent flow. The farther from the front of the car this occurs, the lower the drag on the body, for that means larger areas with low-drag laminar flow, and smaller areas with turbulence. The separation is inevitable; only the exact point at which it occurs varies in accordance with body shape, speed, and air density.

But not all boundary separations are equally turbulent. In a smooth transition, an innermost layer still remains. But when the separation is total, the entire boundary layer rushes into a vortex region where the air generates a suction-type of drag against the body. As the outside airflow exerts a pull on the vortices, there is a buildup of increased negative pressure in the vortex region, and this negative pressure at the back end of the car spells higher drag.

On a proportional basis, with aerodynamic drag as 100, form drag accounts for 55 percent, interference drag for 17 percent, internal flow drag for 12 percent, surface drag for 9 percent and lift drag for 7 percent, according to a General Motors paper dating from 1965. The figures are, of course, not absolute, but reflect typical conditions with a typical GM product. Let us accept them as generally valid and consider each drag factor in turn.

Form Drag Produced by Body Shape

Form drag is due to the body shape. Cars of different shape have different form drag because some shapes let the air pass more smoothly over them. Others cause the airflow to break away and form eddies or vortices. The body shape determines the distribution of aerodynamic pressure over its

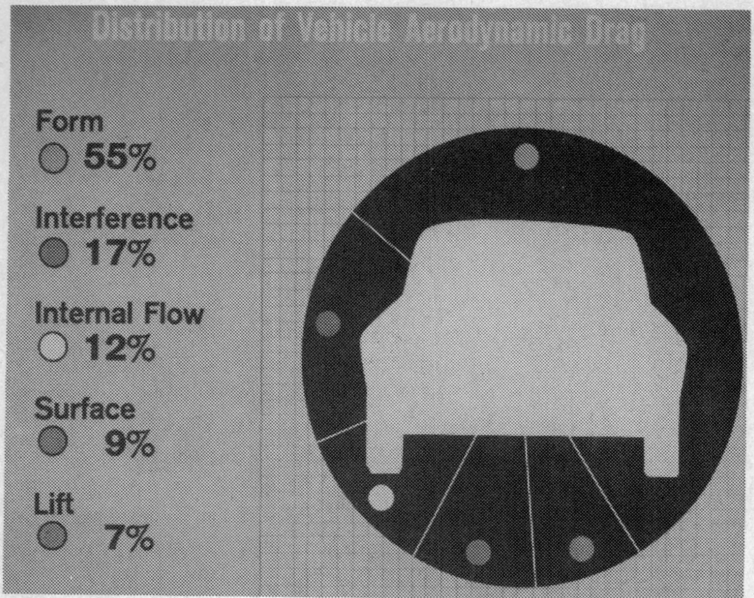

Form drag is by far the main component of aerodynamic drag on a body, but other types of drag add up to nearly half the total. Internal flow drag is the least studied area and perhaps the easiest to improve.

entire outer surface—grille, hood, roof, windows, doors, fenders, trunk lid. Basically, the car with the lowest form drag is the car that presents the surface with the lowest aerodynamic pressure on the front and the lowest "vacuum" or suction forces (negative pressure) on the rear end.

A well-known aerodynamicist, J. J. Cornish of Lockheed-Georgia, has stated that the ideal proportions for low drag is a body 2 1/2 times longer than its width. If the body is longer, drag grows because of the increase in area touched by the air. If it is shorter, there is greater pressure drag due to air flow separation.

As part of the form drag problem, we must include the air drag caused by the wheels. This is called dynamometer loss, because it is due to the wheels wasting rotational energy by spinning the air around them. Each wheel is like a centrifugal compressor rotor—fortunately a very inefficient one.

Lift Drag Produced by Air Under Car

Aerodynamic lift forces occur as a result of a difference in the air pressure acting on the underside of the body and that acting on its upper surface.

A car body has large topside areas where negative pressure is exerted, while pressure may build up under the car.

Apart from the traction and stability problems associated with excessive lift, higher lift forces mean higher drag. That's because the power to generate the lift must be provided by the engine. For two cars of the same power, the one with the higher-lift body will have lower top speed.

Lift distribution is more important for controllability than for drag reduction. It is the lift distribution that determines the pitching moment—a nodding motion in the body. Lift distribution is usually measured at the wheels, and expressed as front lift vs. rear lift, but the actual pattern of lift distribution is usually quite complex, without clear lines.

The aerodynamic lift characteristics of a car are particularly dependent on the design of the lower portion of the nose. That's because the greater the air flow mass below the car, the greater the lift problems.

Surface Drag Produced by Gaps and Trim

You will probably grasp the idea of surface drag best if you think of it as "skin friction." It is a boundary layer phenomenon. The energy lost to skin drag is converted into heat, which is dispersed into the boundary layer.

It is true of all mass-produced cars that the designers have found it impossible to keep the surface completely smooth, due to such elements as windshield frames and recessed windows. It is hard to imagine today's drivers tolerating fixed windows, and nobody has found a way to open and close windows that fit flush with the body surface.

Most of the exterior brightwork and trim on a car is immersed in the boundary layer and presents surface irregularities that count in total surface drag. Gaps between body panels and other imperfections in body assembly count rather heavily. Whether the paint surface is smooth or rough has no practical importance. But the size and shape of body fasteners can be important.

Any and all of these imperfections can cause boundary layer separation. In the best cases, laminar flow is restored after a short distance. In the worst cases, laminar flow is lost forever.

Interference Drag Comes from Add-ons

Anything that projects from a body sets up its own aerodynamic disturbance. Outside mirrors, hood ornaments, windshield wipers, door handles, radio antennas, and rain gutters all add to the total air drag. The losses they cause are called "interference" drag, because these items are added to the basic body shape.

All protuberances on the body surface cause a disproportionate rise in drag, which is due to the accelerated airflow over the body. Reducing the individual drag of such items is the most direct approach to the problem. For instance, outside mirrors should be faired-in to present a bullet-like nose for easier air penetration.

Interference drag can be minimized by grouping such items together, and placing them in areas of the lowest possible air speed and pressure. Exterior hardware of good aerodynamic shape, however, should be spaced out as far as possible. Streamlined door handles are an example.

Interference drag is relatively unimportant on a car with high total drag. If the basic shape is bad, it does not matter if the exterior hardware is badly designed and positioned. But on low-drag bodies, it can be very important. The smoother the basic body shape, the worse the effect of little things that disturb the air flow over its surface.

Tuft study provides "live" visualization of air flow. Sloping hood is favorable for laminar flow, and rounded fenders are helpful, too. This model was one of many Ford tested while developing the Falcon.

Internal Flow Drag

An internal air flow is necessary for the well-being of the car occupants as well as for engine breathing and cooling. The first rule must be to avoid taking in more air than is required. The second rule follows automatically and says the body must provide minimal restrictions on internal air flow and easy escape.

Radiators for water-cooled engines are often larger than they need be, and the air intakes often have an exaggerated frontal area. If correctly placed, a small opening can supply all the air needed for the radiator, for brake cooling, and for engine breathing.

What happens to the air once it has passed through the radiator is a question many engineers seem to have overlooked. Ideally, the air should be ducted outside as soon as possible, with an outlet in a negative-pressure area for maximum extraction effect. Brake cooling air usually gets caught up in the turbulence of the rotating parts and has to find its own escape paths. There are no reliable figures on the importance of the drag that is caused this way, but all engineers accept it, for brake cooling is a point of far greater significance.

The air that enters the air cleaner is consumed by the engine and escapes via the exhaust system. It is advantageous from the engine's viewpoint to get cool air for maximum volumetric efficiency, and in most cases a reasonable compromise with aerodynamic efficiency can be found.

Ventilation air for the interior can be a source of needless increases in air drag. The fresh air intake should be positioned

Not a futuristic study but a concrete proposal for a streamlined four-seater coupe by Pininfarina, the Cr-25 is built on a Ferrari chassis. Slits in hood extract the cooling air. This body also abounds in safety features.

to ingest only clean air, which indicates a high-level location. But the cowl area is not necessarily the most efficient, for turbulence may reduce the flow mass available.

Outlets of ample capacity to prevent pressurization of the interior should be provided, and located in negative-pressure areas. Proper ventilation cannot be assured unless the body engineers have maps of the pressure distribution over the whole body at all speeds and with a variety of yaw angles. Finally, it is important to realize that internal airflow influences not only drag but can have substantial effects on lift and pitching.

chapter 2
the wind tunnel

Some car manufacturers claim an advantage for their products because they have been "wind-tunnel tested." This is, of course, fallacious and misleading. The wind tunnel is simply a tool which cannot originate or develop improvements on its own. All it can do is measure differences in the air flow characteristics of bodies of different shape.

That is very important, however, and even more important is the fact that the wind tunnel is the only form of aerodynamic test facility in existence. The alternative is to test a real car on a real road, and that poses insurmountable problems. You can't control the environment; the instrumentation is both expensive and cumbersome; and the calculations involved are difficult.

The wind tunnel is a test instrument made to measure four different aspects of aerodynamics: drag, lift, pressure distribution, stability; and to provide air flow visualization.

How the scientists can read all that data from measurements of air flow over a body and the forces acting on the body is due to something called a Reynolds number. It is the basis for all wind tunnel testing, as an expression of the principle of dynamic similarity. It is a dimensionless quantity that takes account of flow velocity, air density, and the viscosity coefficient of the air.

The Reynolds number is different from the drag coefficient in that it does not apply to a body but to a specific

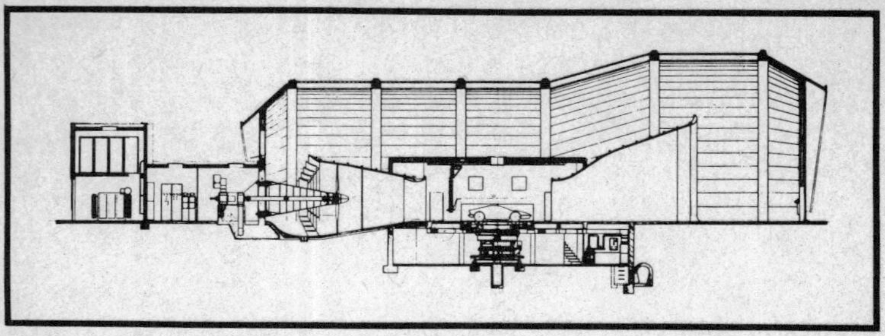

Side view of Pininfarina's full-scale wind tunnel shows the huge size needed to give true readings. If car blocks too large an area of the test section, results are often invalidated. Below the car, you can see the six-component scale and its measurement equipment.

air flow. Low Reynolds numbers mean that the flow is laminar (streamlined), while high Reynolds numbers indicate turbulence. There is a transition range from about 2,000 to about 3,000 when the flow changes from laminar to a wake. Below that it is laminar, and above that, turbulent.

Wind Tunnels Simulate

What goes on in the wind tunnel is a laboratory simulation, not a real-life test. Even with a full-scale car instead of a smaller model, the true outside conditions cannot be faithfully reproduced. The key to wind-tunnel testing lies in the term I have used earlier, namely "relative air flow." Once you accept the notion that it's immaterial what moves and what doesn't, as long as the relativity is preserved, it follows that you'll agree to the precept that instead of driving the car through stationary air, you can just as well blow air over a stationary car.

The next question is how to produce an airflow that's a reasonable facsimile of the air a moving car meets on its way? Obviously, you can't pump it in by a giant bellows or any other device with intermittent action. You need some sort of continuous-flow pump, and what has proven to be the most suitable is a fan. But what about prop-wash? That would produce a highly turbulent wake, not at all the kind of airflow that exists around an actual car body.

Well, there are two basic ways to eliminate prop-wash. One is the Eiffel method. Here, the wake from the fan is dissipated into the outside air. The wind-tunnel test section is placed on the suction side of the fan, where fresh air rushes in to fill the vacuum produced by pumping the air from inside to outside.

The other is the recirculation method. Here, the wake from the fan is collected in a duct, redirected by a couple of right-angle turns, straightened by rectifiers, and finally aimed at the car through a diffuser. Behind the test section, the air is collected and recirculated.

I believe both methods have equal merit, but the experts tend to prefer one over the other. Of recently constructed automotive wind tunnels, only Nissan's in Japan is of the Eiffel type, while most others are variations on the basic recirculation type. Basically, the Eiffel type costs less to build, but the recirculation type costs less to operate (consumes less energy). All modern wind tunnels have electrically driven fans.

Making Air Flow Visible

There are several ways to make air flow patterns visible. First, there is the "wet paint" method, which consists in

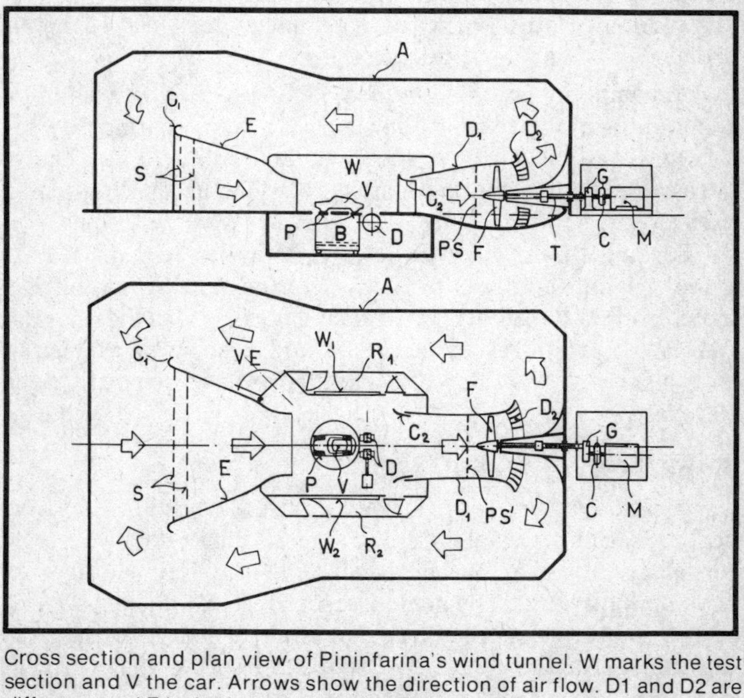

Cross section and plan view of Pininfarina's wind tunnel. W marks the test section and V the car. Arrows show the direction of air flow. D1 and D2 are diffusers, and E is the funnel leading to the test section.

Air flow visualization by lamp black on the body surface at Chrysler Corporation. Fastback scale model shows laminar flow over entire roofline and smooth transition over rear quarter panels and fenders.

mixing lamp black and kerosene, coating the model or complete car with this mixture, and then blowing air over it. The pigment (lamp black) spreads under the speed of the air and traces the air flow paths on the surface.

A variation on this method is to use a lightweight oil containing a fluorescent powder. When a coating of this substance is spread over the body and photographed under ultra-violet light, the flow paths will show up as illuminated traces.

Second, there is the tuft study. A multitude of yarn tufts glued—at one end—to the body in a certain formation can indicate conditions beyond the boundary layer, as the tufts are free to rise from the surface. Tuft studies show air flow paths on the surface, and also illustrate patterns of separation and turbulence.

Tunnel Testing with Scale Models

Nowadays, automotive wind tunnels are full-scale, using real cars. But the whole business began with scale models, and model tests can still be valuable in the intiail stages of developing a body. The scale is usually 1/10, 3/8 or 1/4 actual size. It is best to use the largest possible model, because small differences in the bodies have little effect on the Reynolds numbers. When tests are made with small scale models, the results are not totally reliable. This deviation is known as "scale effect."

It is due to air flow patterns over the model being different from those over the full-scale body. Whenever there is a large scale effect, the model tests will give a misleading forecast of the body's drag coefficient.

F. N. Beauvais of Ford Motor Co. claims, in a 1969 report, that it is desirable to use the largest possible model. On the other hand, says Beauvais, the model should be relatively small in relation to the tunnel test section. The ratio of model frontal area to tunnel cross-sectional area is called "blockage ratio." A blockage ratio of 5 to 7 percent is considered ideal, and if it exceeds 25 percent, the test results are no longer trustworthy. Some experts disagree, however.

Professor H. Schlichting of the Institute of Fluid Mechanics of the Technical University of Braunschweig in Germany feels that the results of wind tunnel tests on automobile bodies do not depend too much on the size of the model or the size of the wind tunnel. He claimed in a paper presented in 1960 that even very small wind tunnels with a

Scale-model testing is seldom accurate, but gives an indication of whether the basic shape is sound or not. Costs are far lower than in full-scale testing.

working section of 3 to 5 feet diameter could be used for evaluating aerodynamic problems.

Ground Effect on Scale Models

A problem that stems directly from the simulated environment in the wind tunnel is what is known as ground effect. In real life, the car is moving relative to both the air and to the ground. On the open road, the air moves relative to the car but not to the road. There is no boundary layer on the roadway. This is one thing the wind tunnel cannot simulate, because in the tunnel, the wind moves relative to the ground and model alike, but the model or car has no motion relative to the roadway.

In full-scale testing, the ground effect problem is considered minimal, but aerodynamicists have grappled with its influence on scale-model testing from the very beginning. Aircraft designers suspended their models in the free air flow, and early tests on car bodies were made the same way. This method, of course, simply ignored the basic difference between automobiles and flying machines. The next step was to use two models, mounted as mirror images, to simulate ground effect. The theory was that the identical models would provide a plane of symmetry in the air flow between them, but practice has shown that this is not in fact what occurs.

This led to experiments with various forms of floor mounting, which is now universal for automotive wind tunnel

Tuft study on Ford Falcon prototype shows vortex formation at windshield pillars and in rear window area. The tail shape gives rise to massive turbulence in the deck lid edge and bumper area.

testing. Some scientists prefer a fixed ground board to a mounting that's flush with the surface, however. They argue that the ground board simulates a roadway better, because it gives rise to a separate boundary layer that remains thin for the operative length of the measurements. With flush-floor mounting, the boundary layer may build up to disproportionate thickness and invalidate the test.

Other aerodynamicists say that technically accurate results can only be obtained with a moving belt, geared to run at air speed. This has not been proved, however, and its construction introduces mechanical problems that have not been overcome.

Wheel rotation has not been possible to simulate, with either models or full-scale cars. Its effect on drag may be relatively small, but as Donald S. Gross and William S. Sekscienski of the wind tunnel operations department at the University of Maryland have pointed out, "More data should be available before wheel rotation is dismissed as unnecessary. It is possible that wheel rotation may produce larger effects on lift and pitching moment than on drag."

Measuring What the Tunnel Tells

Measurement techniques have been developed that provide a nearly complete map of the body's airflow characteristics. The readings are not taken directly on the model, but from a balance or scale place below it. The best balances are of the six-component type, which does not mean that they have six parts, but that they are able to measure six

Full-scale mockup goes into the wind tunnel for testing. Pontoons are substitute for six-component scale, measuring both vertical and horizontal forces.

Road-test prototype is used to test cooling and ventilation systems under real-life conditions. Main air intake is relocated below bumper, while grille serves primarily cosmetic purposes.

different inputs. They are drag, lift, side force, pitch moment, roll moment, and yaw moment.

The balance is usually built up as a pyramid, with a balance strut extending upwards from a main balance beam located deep in the support system. With this type of construction, you can measure horizontal as well as vertical forces.

Sidewinds are simulated by simply turning the platform to one side, to change the car's heading relative to the air flow. Underneath the surface of the platform or turntable which carries the model, you find an H-frame hoist with individual

Final product embodies lessons learned from both highway and wind tunnel testing. Ribbing along door sills and rocker panels is intended to keep road splash dirt from creeping up on body sides. Door handles are streamlined, but window frames are not.

wheel supports. Holes in the platform surface allow the model wheels to rest directly on these wheel supports. The same techniques are applied in full-scale wind tunnels, with giant industrial balances instead of little desk-top rigs.

Pressure distribution patterns are usually established by inserting small tubes in the car or model surface (1/16 to 1/8 inch diameter) and reading the pressure on a multiple manometer or an electrical pressure transducer. With the manometer, the tubes are simply connected to the instrument via plastic tubing. With the transducer, the tubes are connected to scanning valves which sequentially link each tube to the transducer, which sends a signal to an electronic pressure gauge. The second method involves more expensive equipment, but can handle higher numbers of tubes.

chapter 3
aerodynamics and car design

Low drag is the primary consideration in adapting streamlining to mass-produced cars. Other aerodynamic phenomena, such as lift, are not basically associated with everyday cars but with more esoteric machinery used in various forms of racing. I will discuss those matters in a later chapter.

The design elements under consideration at the moment are those that play a part in the streamlining of the average car and can help improve its fuel-efficiency. It's no secret that in order to meet the government's fuel economy standards for 1980 and 1984, the industry will find itself bound to improving the aerodynamics of its products.

There are many practical considerations which have prevented aerodynamics from playing a greater part in the design of car bodies. First of all, the car's duties are to carry people and goods. A family car must be designed from inside out, not from the outside in. That means you must start with the seats and the required legroom and headroom, then add the power train, trunk, fuel tank, spare wheel, and accessories. Finally comes the body shape, more or less as an envelope for the whole. You cannot start with a spacecraft shape and then fit the mechanical components and passenger compartment into it.

Sports cars, with reduced needs for carrying capacity, can be brought closer to ideal aerodynamic shapes than family

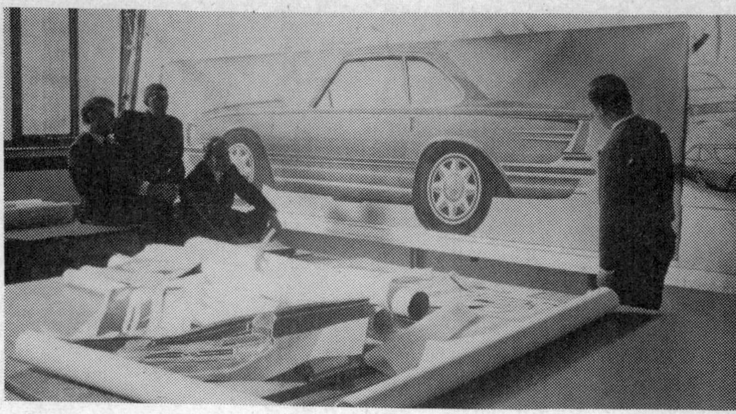

Basic shape of the Mercedes-Benz 450 SL was born on the drawing board, in full scale. Three-dimensional detail sketches crowd table and walls while designers discuss ideas for improvement.

cars. Special-purpose vehicles, for road racing, oval-track racing, and even straight-line speed records, can be brought much closer to conformity with spacecraft concepts, but even they still have wheels that must be housed.

Nagging Requirements Spoil the Ideal

There are other design constraints on all cars designed for use on normal highways. They must fit dimensionally into the traffic picture. Their width must not exceed 80 inches. They must be short enough to negotiate street corners. The driver must be positioned so he has clear vision. Places must be found for lamps, license plates, etc. The designer of any mass-produced sports car or family car has an extremely difficult task, in that he must reconcile his overall concept of shape with a long list of design elements, more and more of which are subject to legislation.

Naturally, the car body is not assembled from a multitude of individual design elements, but conceived as a whole, as one shape. Nevertheless, there are many design elements that must be kept constantly in mind while this shape is being fashioned. Therefore, it becomes indispensable to consider the various elements separately in order to draw maximum benefits from each in developing the overall shape.

Nose design is the most important aspect of body shape, for it influences the whole airflow picture around the entire car. It is a main factor in reducing drag, and it plays a major part in pitch control.

First, the nose must be "soft." That does not mean structurally soft, like an energy-absorbing bumper system. It means soft in an aerodynamic sense—an absence of brutal fronts, vertical walls and sharp corners. No matter what part of the car you are talking about, it remains true that air cannot flow smoothly around sharp corners. Some compromise between fashion and function has to be made, and fortunately, fashion can be molded to conform to functional needs.

Just Four Basic Shapes

As far as the overall shape is concerned, we have to make distinctions between the various types of car. The average family car has a box-on-box shape, that is, a big basic box containing hood and deck lid, with a smaller "greenhouse" box on top. Station wagons are shaped like vans, with a hood in front. Sports cars come in two basic forms—coupe and convertible. The coupe is usually an oval-silhouette shape, whether it has its engine in front or in the back. The convertible, or "roadster", has the overall proportions of a brick, rounded off to resemble a wedge, a teardrop, or a coke-bottle.

To minimize the frontal area, the body should be no wider than necessary to cover all mechanical and load-carrying elements. The advantages of a low roofline are obvious. Not only does it bring about a direct saving in frontal area, but it makes streamlining easier. Tangential streamlines (laminar-flow characteristics) are more easily obtained when the overall height is kept down.

The front wheels steer the car and must have room to turn. If the front wheels are to be enclosed, the designer has the choice of either a wider body, a narrower track, or reduced steering angles. None is practical for normal "street" cars.

The front must have a low aspect. That's essential. Remember the cars of the 1950s, with their low, wide grilles, topped by a huge dome of sheet metal forming part of the hood? That was bad. Flattening the hood to meet the top of the grille was a big help, but many grilles remained too aggressive to the airflow, and caught far too great a volume of air which had to be pushed through the engine compartment.

Where to put the headlamps is a question that few designers are comfortable with. American law makes problems by forbidding Plexiglass covering for headlamps,

Modulo by Pininfarina combines a variant on the Goettinger form with the idea of making the car symmetrical around a horizontal plane located about wheel hub level. The chassis is a midships-engined Ferrari.

but the fashion for dual headlights created almost as many problems. Generally, from a streamlining viewpoint, it is always better to put the lamps side by side than to stack them vertically. Tall light units make for tall fenders, and they can cause a lot of drag.

Splitting the Air Four Ways

As the car comes down the road the air flow separates in front of the vehicle along a "stagnation line." It is usually located about bumper height. All the air above it is deflected over the car, and the air under the stagnation line passes beneath the car. There is also a lateral stagnation line, where the air mass is split right and left. A projecting central point tends to cause less disturbance and confusion in the air than twin projections on the sides.

How pointed the nose should be is a matter that must be considered individually for each car design, with due regard to its speed range, operating conditions, length and weight restrictions, bumper standards and body-impact absorption characteristics.

The hood must form a transition between the nose and the windshield. On many prototypes you find there is no hood as such. The hood line goes straight from the nose into the windshield, as if you had taken the front of a bus and tilted it back until you reached sports-car roof height.

Hood length? There are no hard and fast rules on that. Many designers wish they could do away with it altogether, but it's funny what happens when there's no reason to have one.

After many years of a flat-nosed Econoline, Ford put a hood on it a couple of years ago. In aircraft, Sud-Aviation's Caravelle eliminated the hood, but the more recent Douglas DC-9 brought it back.

Ideal hood length really depends on the rest of the body. On the Datsun Z-car, you obviously need a long hood. On the Fiat 128 it clearly has to be very short. Hood length depends to some extent on its slope, which is in turn dictated by cowl height and nose shape. All of these elements must be blended into a harmonious common form, and that's where the real artistry comes in.

Hood-to-windshield a Critical Point

The hood must blend neatly into the fenders and cowl. Depressed hood designs usually cause higher drag. Bulbous hoods, which tend to be aesthetically offensive, are often aerodynamically advantageous. At the windshield base, the air flow is forced to a new and higher level, and two things must be considered: windshield slope and windshield section.

A steep (i.e., near vertical) windshield will set up a disturbance in the air flow well ahead of its location. Just as the air flow separates at a stagnation line in front of the car, the windshield presents a similar obstacle which gives rise to stagnation-separation in the air flow over the hood of the car. What happens here is that an energy-consuming vortex formation is caused when the boundary layer on the hood moves into the high-pressure area at the windshield base.

The windshield should be well rounded into the side windows to avoid airflow separation at the side posts. The middle area can be straight if the corners are smoothly

Bertone eliminated the hood as a horizontal panel and made it an extension of the windshield on the Lamborghini Bravo. V8 engine is midships-mounted.

curved. Experiments have shown that there is nothing to gain by curving the entire windshield into a blunt C-front. And a split V-front windshield usually has higher drag than a flat windshield.

A smoother transition from the base to the roof is of primary importance. The flatter the windshield rake, the easier these transitions will be. But there are practical limits here. Visibility from the driver's seat demands a certain height, and with front-mounted engines, there is limited length available because the hood must open wide enough to permit engine removal.

There are visibility problems associated with increasing the distance between the windshield and the driver's eyes that argue in favor of putting the windshield base in close proximity to the steering wheel.

The Roof Problems—and the Fastback Fallacy

Roof design, too, is three-dimensional. It's important to get the widest point in the plan view well forward, and to give it a cross-section that prevents vortice formation around the edges. But the worst problem is the profile.

The body designer's big challenge is to make the boundary layer stick to the roof of the car. Since air tends to flow from a high pressure area to a low pressure area, a gentle taper towards the rear has been found to work favorably towards the retention of laminar flow.

But the kind of taper that gives the best boundary-layer control, would also produce an impractically long body. What the stylists and advertising copy writers call a "fastback" is not a good idea for lowering the air drag. The roofline slopes too abruptly to retain laminar flow, no matter how smooth the transition over the rear window and trunk lid.

The roofline should continue to rise at the top of the windshield in order to ensure a smooth transition, and then level off in a very gentle arc. At some point this arc must end, to be followed by a straight roofline. It need not be absolutely horizontal, but it cannot exceed about 10 degrees downwards slope from horizontal without causing separation.

Lower drag can better be achieved by a long, straight, near-horizontal roofline, ending in a sharp edge, with a near-vertical rear window (as on station wagons), rather than by the fastback concept, which would need two car-lengths to

You'd never guess it, but the boxy Ford Econoline was developed in the wind tunnel. Scale model is an exact reproduction, with door handles, door hinges, window frames and lamps faithfully rendered, to minimize errors.

be effective, and is therefore impractical for cars intended for everyday use.

Turbulence at the tail end is impossible to avoid. How tail lights, trunk lid handles, and rear bumpers are shaped will not have a large influence on total drag. What's important is to minimize the area of turbulence. And that comes down to retention of laminar flow on both roof and sides as far back as possible. With a neatly tapered body shell, the turbulent area can be reduced to perhaps 2/3 of the car's frontal area.

Enclosing the rear wheels will help reduce drag in the critical area where the air flow along the body sides adopts its final pattern prior to merging with the roofline flow and the air from below the car.

The Complication of Wheels

The question of wheels is complicated by the fact that they must be placed wide apart, and preferably near the four corners of the body, for the best controllability, stability, and ride. They cannot be hidden deep in the middle. The increase in form drag by the addition of wheels to the body is readily apparent. But the effect of the wheels on air flow are not so easily seen or understood. The basic problem is not the size and shape of the wheels, but the fact that they run on the road, which forces the car body to move in a layer of air that's closest to the ground.

The air drag of the tire itself varies with its exposed area—the amount of tread that shows up as frontal area. There is a theoretical advantage here in choosing narrower tires, but that's hardly a practical solution. Most production cars would not be safe with narrower tires.

It is inevitable that air from below the car should also mix with the tail-end turbulence. The air that is diverted under the car flows under different conditions than the air around the body. Below the car, the air is in constant shear between the stationary roadway and the moving bottom of the car. Consequently, this air does not behave in the same way as the air flow along the sides and over the roof.

Underneath—Still an Area of Dispute

Many scientists maintain that the proximity to the ground tends to diminish the importance of streamlining the underside of the car. Most car manufacturers have been content to leave it out of sight and out of mind. There is little reliable knowledge of the subject, and more data is needed before the industry can justifiably dismiss the question as one of minor importance.

On the other hand, there are aerodynamicists who argue: When you put the car body closer to the ground, the air flow between the chassis and the road surface is rapidly speeded up. That introduces the risk of increasing the drag due to higher energy losses over the rough lower side of the car. As a result, drag may be lower when the ground clearance is higher. The clearance that gives maximum drag differs from car to car, but all modern mass-production cars are fairly close to it. The differences between individual models is due to differences in overall design, and in the thickness of the underside boundary layer.

One reason why this is still a relatively uncharted area is the fact that turbulence in the back-bumper area is not necessarily caused by the lack of a smooth undertray. It can stem from vortex formation along the fender tips, flow separation in the decklid area, and interference from the rear wheels.

Adding a smooth skin under the chassis will not have any beneficial effect on overall air drag if such conditions prevail.

On the other hand, if you have a body that would retain laminar airflow right to the end, if it had not been disturbed by vortex formation in the air escaping from underneath the car, there, a smooth underpan could make an important contribution.

The Neglected Interior Aerodynamics

One design element remains to be discussed. I am tackling it last because it is probably the most neglected. I am referring to internal air flow. To start at the beginning, cars have internal air flow for several reasons. First, the engine needs fresh air (oxygen) for the combustion process.

Many mass-production cars let the carburetor breathe through an air cleaner located in the hottest area under the hood. Fortunately, it is getting more and more common to provide the air cleaner with "horns" that suck cooler air from points further forward. That is only a halfway step.

For efficient feeding of air to the engine, outside air should be taken in at a high-pressure area, such as in the side of the radiator air intake, or through a separate air scoop. Fuel-injected Mercedes-Benz and other high-technology cars were first to adopt such systems.

Ram-air induction is often a fallacy. The forward-facing air scoop on the hood, feeding into the air cleaner, is totally ineffective at low speeds. Even at high speeds, its benefit is dubious. Reporting in 1963, Walter Korff of the Lockheed-California Company described the ram effect at 75

Corvette from 1973 has good streamlining for a front-engine car with a long hood. Tail end design, however, does nothing to retain laminar flow. Internal air flow is well controlled.

to 100 mph as "slight" and accounting for a horsepower gain of only about one percent.

Backwards-facing air scoops, drawing in high-pressure air dammed up in front of the windshield work better on many cars. But when they do work well, it is only testimony to the poor airflow characteristics of the car body. If smooth, laminar flow could be maintained across the hood-cowl-windshield juncture, the total drag would be lower, and the pressure would not be high enough to give valuable assistance to the air scoop.

Most of the air that enters the grille is ducted through the radiator. Airflow through the radiator approaches but is not equal to the speed of the car. As the air enters the intake, it is slowed down in accordance with the ratio of opening size to radiator size. This reduction in air speed is accompanied by a build-up of air pressure, which is most useful in forcing the air through the radiator.

It should be noted here that ram effect, perhaps somewhat surprisingly, plays no big part in this. But the pressure gain produced by the slowing-down of the air can easily be lost if there are leakage paths between the air intake and the radiator. To be effective, the pressure gain must be protected by tightly sealed ducts. Few production cars take advantage of this, as they leave a long open space between the grille and the radiator, open to both sides and to the ground below.

Behind the radiator, the air is let loose in the engine compartment, where it flows around the engine and its accessories, and often hammers out its energy against the cowl. Air flow over the engine block and head is not important. With a properly engineered cooling system, it is more efficient to cool the block and head by cooling the water that circulates through them.

Ducting the air away from the radiator can be equally important. The air flow inside the engine compartment is subject to the same skin friction, interference drag and turbulence as the air flowing around the car.

A full duct is perhaps not practical, due to the need for service and maintenance accessibility to the engine. But it is certainly possible to provide a low-restriction flow path for the air from the radiator, and to position its outlet in a convenient low-pressure area which will then provide an extractor effect. If the outlet is placed in a high-pressure area, the escaping

Four-rotor, experimental Corvette has concealed headlights which pop up from hood panels when switched on. Window frames have been designed for flush mounting of the glass, but you look in vain for an escape route for the cooling air.

cooling air will be fighting the external air flow, adding to the total drag. An example will illustrate how difficult it is to properly control underhood airflow.

When J. J. Cornish was working in aerophysics research at the Mississippi State University in 1963, he published an analysis of the airflow around the MGA roadster. This car was the first "streamlined" MG, replacing the traditional-style TF. But it had not been wind-tunnel tested before it was cleared for production. Its air intakes and outlets had been positioned partly by guesswork and partly by mathematical calculation. Sheer discretion forces me to silence as to which played the greater part.

Anyway, the car had two vents ahead of the cowl, in the valleys between the hood and the fenders. These were intended to extract hot air from the engine compartment. That would indeed have worked if the vents had been located in a low-pressure (negative-pressure) area. But due to stagnation-separation ahead of the windshield, it was a very high-pressure area, with active turbulence. The vents adopted an inward airflow. Instead of serving as extractors, they became intakes. Under the hood, the air flowed forwards, to fight the cooling air coming through the radiator.

Air for brake cooling should also be taken in separately. This is really an external function, for the brakes are exposed parts, cooled by the air that flows under the car.

The important thing to realize about aerodynamics in respect of brakes, is that it is possible to make use of aerodynamic design to obtain better brake cooling.

chapter 4
how to judge streamlining

It would be nice, for you who have no wind tunnel, if you could look at a car and rapidly estimate its drag coefficient. Obviously, you can guess whether a car has high or low drag, by looking at its basic shape. But when you have two cars that are closely similar, can you really tell which has the better aerodynamics? Maybe you can't, but you can learn. R. G. S. White of the British Motor Industry Research Association test establishment has developed a method of estimating the drag coefficient, and it works with remarkable accuracy.

He divides the body shape into six basic zones. They are nose, windshield-to-roof junction, main roof panel, rear roof-to-deck-lid, lower tail end, and underbody. Each zone is examined three dimensionally and given a rating on the basis of known airflow characteristics, as interpreted from MIRA's bank of full-scale wind tunnel tests.

Nose

The shape with the highest drag is the box—a squared-off front section with a vertical face, high and horizontal hood line, constant width, and sharp corners. A graduated scale indicates the effect of step-by-step improvements. Fenders with tapered-in square corners are better than the full-width box shape. Rounded corners are better still, and rounded corners without protuberances are better than rounded corners with protuberances. A lowered front with sloping hood is better than a high, rounded front horizontal hood line.

Corvette prototype with midships-mounted rotary engine has clever nose design with front spoiler, but headlamps recessed into NACA ducts act as buffers to air flow.

Well-rounded front fenders bring still further improvement, and in this combination, a low, square front with a sloping hood again comes off better than a high tapered hood line. Best rating goes to the nose that approaches a semi-circular plan view, like a snub-nosed bullet, with a sloping hood.

Windshield-to-roof Junction

Design elements making for highest drag in this zone are flat windshield with sharp corners, near-vertical rake, and a depressed hood between high square-edged fenders. This sounds like a description of the Renault 8 and Simca-1000. A hood flush with square-edged fenders is to be preferred to the central depression in the hood.

It would be better to have rounded fenders leading to a high cowl, and that can be improved by lowering the hood to fender level, and giving the windshield a convex curve. For lowest drag, look for well-rounded windshield corners, flush hood and fenders, with low cowl height, and well-rounded body sides.

Main Roof Panel

Any forward-projecting peak such as an exterior sun visor, will raise drag to new heights. A reverse taper on the roof in plan view, so that it's widest at the rear and narrowest at the front, will substantially increase form drag. A squared-edge windshield flange, with protruding rain gutters along the sides of the roof, are grounds for serious demerits. A rounded front edge, where the roof meets the windshield, is a prerequisite of low drag. A passenger compartment with symmetrical layout, widest at the middle and tapering towards both front and rear ends, with a TV-screen outline for its roof in plan view produces higher drag than a shape that is widest at the front and tapers off to the rear.

Roofline

A rounded-off roof section, with short deck, or no deck, tends to produce very high drag. That describes the original Mini. Slightly better is a sharp-edged roof, with sloping rear window, and short (or no) deck. This is a popular style for modern small cars (VW Rabbit, Fiat 127, Volvo 343), and as a shape it is less favorable than the combination of a rounded roof and rear window, with a rounded trunk—along the lines of the old Hillman Minx.

Something to be avoided next, according to White, is an outward taper in the rear quarters, fenders, or flared-out tailfins. The boxy sedan body, with little-box-on-big-box proportions, does not fare so badly. Square design of this area, with sharp corners, is better than rounded. Fiat 124 and 131 do not look streamlined, but their rear roof and deck shapes are quite efficient.

A small taper towards the tail, in the whole body, is a point in favor. Increased taper increases the advantage. A roofline that slopes to a cutoff point about deck level is better than a slope that continues to bumper level. Best rating goes to the roofline that maintains a gentle, continuous slope right to the tail, and is cut off at a high level. Cars like the Alfasud and Citroen GS and CX exemplify this school of design.

Underbody

Worst rating goes to the deep-chassis car with a maximum of mechanical parts projecting into the air stream (suspension arms and springs, exhaust system, steering linkage, drive shafts and axles). Any form of cover, however partial, gives a drag reduction. Platform frames (VW Beetle, Porsche 911) go a long way towards underbody streamlining. An integral, flush floor pan with little projecting mechanism, would be ideal.

Drag Coefficients by the White/MIRA System

From points given in the drag rating, White calculated the drag coefficient by a simple formula of dividing by a constant to correlate the numbers with those actually obtained in full-scale wind-tunnel testing.

The maximum scatter about the mean line proved to be plus or minus 7 percent. For most cars, his calculated data are correct to the second decimal, and for some even to the third. For instance, he rated the 1964 Ford Galaxie at a C_D of 0.436, which coincided perfectly with the test data. The 1966 Oldsmobile Toronado was calculated to have a C_D of 0.377, while tests showed it to be 0.380.

His error on the 1965 Mustang was slightly larger, with a calculated figure of 0.454 and a measured figure of 0.475. His rating of the VW Beetle was extremely close, at a C_D value of 0.445 against a test value of 0.448. On a car as difficult to judge as the Saab-96, he erred slightly in favor of the designers, with a calculated C_D of 0.341, as compared with a measured C_D of 0.358.

On superbly streamlined cars like the Citroen DS-19 and the Porsche 911, the White ratings erred only on one third-decimal point, which is too small to consider. He calculated C_D values for the DS-19 at 0.312 and the 911 at 0.303, while the tests showed the actual values to be 0.311 and 0.302. The error was greater in the case of a boxy shape like the 1963

Porsche 911 has a thoroughly streamlined shape, with a variety of special aerodynamic aids, such as a partial air dam (front spoiler), NACA ducts on the hood, flush bumper ends, and concealed gas filler cap.

Porsche Turbo has more extreme aerodynamics, where anti-lift features play a major part. Front air dam extends closer to ground level to keep air from getting under the car, and rear spoiler relieves negative pressure on the tail. Bulging fenders are not aerodynamic, but necessary to cover wider wheels and tires.

Mercedes-Benz 300 SE, which he calculated to have a C_D of 0.377 while the test value was actually 0.387.

I have used the White/MIRA rating system to calculate the C_D of current cars. In the following table the frontal area has been calculated as 0.81 of the product of maximum width and maximum height, and the total air drag was computed according to the established formula (see page 17 of Chapter 1).

MAKE	MODEL	FRONTAL	C_D	55 MPH Air Drag (lbs)	70 MPH Air Drag (lbs)
Alfa Romeo	Alfetta GT	19.2	0.34	49.4	80.0
Alfa Romeo	Alfasud	17.7	0.33	44.2	71.6
Aston Martin	V8	20.1	0.37	56.2	91.1
Audi	Fox	18.0	0.38	51.7	83.8
Audi	100 LS	20.3	0.39	59.9	97.0
BMW	320	18.4	0.40	55.7	90.2
BMW	530 i	19.9	0.39	58.7	95.1
BMW	630 CS	17.9	0.35	47.4	76.7
Buick	Skyhawk	17.5	0.40	52.9	85.8
Buick	Skylark S/r	20.3	0.45	69.1	111.9
Buick	Riviera	22.6	0.42	71.8	116.3
Cadillac	Seville	21.0	0.40	63.5	102.9
Cadillac	Eldorado	23.1	0.41	71.6	116.0
Chevrolet	Chevette	17.3	0.44	57.6	93.2
Chevrolet	Vega hatchback	18.1	0.40	54.8	88.7
Chevrolet	Monza 2+2	17.5	0.40	52.9	85.8
Chevrolet	Monza Towne Coupe	17.4	0.42	55.3	89.5
Chevrolet	Camaro	19.6	0.36	53.4	86.4
Chevrolet	Corvette	17.7	0.33	44.2	71.6

MAKE	MODEL	FRONTAL	C_D	55 MPH Air Drag (lbs)	70 MPH Air Drag
Chrysler	Cordoba	21.7	0.40	65.6	106.3
Citroen	GS	18.0	0.30	40.8	66.2
Citroen	CX	19.5	0.27	39.8	64.5
Datsun	F-ll coupe	16.5	0.37	46.2	74.7
Datsun	B-210	17.5	0.40	52.9	85.8
Datsun	610 coupe	18.8	0.46	65.4	105.9
Datsun	710 coupe	18.3	0.46	63.7	103.1
Datsun	280-Z	17.4	0.34	44.7	72.5
Dodge	Colt	18.4	0.38	52.9	85.7
Dodge	Aspen coupe	20.7	0.44	68.9	111.6
Ferrari	308 GTB	16.0	0.30	36.3	58.8
Ferrari	365 GT4 2+2	19.5	0.32	47.2	76.5
Ferrari	365 BB	16.7	0.28	35.4	57.3
Fiat	X-1/9	15.3	0.36	41.7	57.5
Fiat	128-3P	16.9	0.32	40.9	66.2
Fiat	131 coupe	18.9	0.42	60.0	97.2
Ford	Fiesta	17.6	0.40	53.2	86.2
Ford	Pinto	18.8	0.44	62.6	101.3
Ford	Mustang Ghia	18.8	0.44	62.6	101.3
Ford	Capri	18.3	0.40	55.4	89.7
Ford	Maverick 2-door	19.9	0.41	61.7	99.9
Gremlin		19.7	0.37	55.1	89.3
Honda	Civic	16.5	0.38	47.4	76.8
Honda	Accord	16.5	0.37	46.2	74.8
Hornet	Hatchback	19.8	0.37	55.4	89.7
International	Scout II	24.6	0.47	87.4	141.6
Jaguar	XJ-Six	20.2	0.37	56.5	91.6
Jaguar	XJ-S	18.7	0.30	42.4	68.8
Jeep	Cherokee	27.0	0.46	93.9	152.1
Lamborghini	Urraco	16.9	0.28	35.8	58.0
Lamborghini	Espada	18.3	0.30	41.5	67.3
Lamborghini	Countach	16.8	0.29	36.8	59.7
Lancia	Beta coupe	17.8	0.35	47.1	76.3
Lancia	Beta HPE	17.9	0.33	44.7	72.4
Lancia	Scorpion	16.7	0.31	39.2	63.4
Lancia	Stratos	16.1	0.29	35.3	57.2
Lotus	Elite	18.2	0.30	41.3	66.9
Lotus	Esprit	17.1	0.29	37.5	60.7
Maserati	Merak	16.5	0.30	37.4	60.6
Maserati	Khamsin	17.0	0.35	45.0	72.9
Maserati	Bora	16.1	0.28	35.2	56.9
Matador	coupe	21.6	0.36	58.8	95.3
Matra-Simca	Bagheera	16.8	0.30	38.1	61.7
Mazda	RX-3	17.8	0.36	48.5	78.5
Mazda	RX-4	19.0	0.35	50.3	81.5
Mazda	Cosmo	18.6	0.36	50.6	82.0
Mercedes-Benz	230	21.1	0.38	60.5	98.2
Merdedes-Benz	450 SE	22.0	0.37	61.6	99.7
Mercedes-Benz	450 SL	19.7	0.35	52.1	84.5

MAKE	MODEL	FRONTAL	C D	55 MPH Air Drag (lbs)	70 MPH Air Drag
Mercury	Comet coupe	19.9	0.40	60.2	97.5
MGB	GT	16.8	0.36	45.7	74.1
Morgan	Plus-8	16.4	0.55	68.2	110.5
NSU	Ro-80	20.6	0.33	51.4	83.3
Oldsmobile	Starfire	17.5	0.40	52.9	85.8
Oldsmobile	Omega coupe	20.8	0.45	70.8	114.7
Oldsmobile	Cutlass S coupe	21.8	0.37	61.0	98.8
Oldsmobile	Toronado	22.7	0.42	72.1	116.8
Opel	Isuzu	17.7	0.43	57.6	93.2
Pacer		21.7	0.36	59.1	95.7
Peugeot	304 sedan	18.5	0.38	53.2	86.1
Peugeot	504 sedan	20.5	0.39	60.5	97.9
Peugeot	604 sedan	21.0	0.39	61.9	100.3
Plymouth	Arrow coupe	18.4	0.37	51.5	83.4
Plymouth	Volare coupe	20.7	0.44	68.9	111.6
Pontiac	Astre coupe	18.1	0.40	54.8	88.7
Pontiac	Sunbird	17.5	0.42	55.6	90.0
Pontiac	Ventura SJ	20.4	0.46	71.0	115.0
Pontiac	Firebird Formula	19.6	0.38	56.3	91.2
Porsche	924	17.7	0.36	48.2	78.1
Porsche	911	17.6	0.34	45.3	73.3
Porsche	Turbo	19.0	0.33	47.4	76.8
Renault	5	17.6	0.39	51.9	84.1
Renault	15	17.7	0.41	54.9	88.9
Renault	30	20.5	0.40	62.0	100.5
Rolls-Royce	Silver Shadow	22.7	0.45	77.3	125.1
Saab	96	19.4	0.36	52.8	85.6
Saab	99 Wagonback	20.2	0.34	51.9	84.1
de Tomaso	Pantera	16.7	0.29	36.6	59.3
Toyota	Corolla Liftback	17.5	0.38	50.3	81.5
Toyota	Celica ST	17.6	0.39	51.9	84.1
Triumph	Spitfire	13.9	0.39	41.0	66.4
Triumph	TR-7	17.7	0.34	45.5	73.7
Volkswagen	Beetle	19.2	0.45	65.3	105.8
Volkswagen	Rabbit	18.8	0.40	56.9	92.1
Volkswagen	Dasher	18.0	0.37	50.4	81.6
Volkswagen	Scirocco	17.6	0.38	50.6	81.9
Volvo	242	20.3	0.35	50.8	87.0
Volvo	246	20.3	0.35	50.8	87.0

Pinto, Mustang, Monza, Pacer, Gremlin

A few comments on some of the cars in the list may be of interest. For instance, Ford Pinto and Mustang have windshields with a gentle curve, that are not much better than straight windshields. Their corners are still quite sharp, and even if the glass is sharply raked, it does not do much to alleviate the problem.

Both have stylized headlight bezels that have a parachute effect on the airflow, in addition to being located right at the fender edges where it is so important to have a smooth, rounded shape. The high vertical grille on the Mustang presents a formidable obstacle to the air flow, and raises the hood lip nearly to cowl height. Shallow gutters between the hood and fenders also make an adverse contribution.

On the Monza 2+2, the benefits of the sloping hood and low grille are negated by the air traps created by the recessed rectangular headlamps. The fender tips are nicely tapered, but the leading edge protrudes. If those lights had been enclosed behind glass, the design would have been good, but without covers, the headlights cause enormous drag. The Vega nose, less drastic, presents a milder face to the wind, with lower air drag.

The front end of the Pacer is exemplary in its concept, with very low grille, smoothly sloping hood, and a coherent cowl structure, but the recessed headlamps disturb the air flow to some extent. The bulbous rear end is curved and short, and a good example of what to avoid. The more sharp-edged Gremlin tail is a more aerodynamically efficient design.

Cutlass, Toronado, Firebird, Grand Prix

The backswept grille of the Cutlass S is certainly not everybody's idea of good styling, but the effect of tilting back the grille is quite important. The front end is slightly marred by air-dam headlamp design and the sharp front fender edges. Along with the Chevelle Laguna S-3, it is one of the better front ends on recent General Motors cars.

The first Toronado, designed in 1964-65, was the cleanest of all models of that name to appear in later years, with a drag coefficient of 0.380. Annual model changes have catered to Oldsmobile's idea of public taste, to the detriment of aerodynamic design. The low grille and concealed headlamps have given way to a blunter nose and "open" lights, while the original fastback has been replaced by a notchback of common design with the Delta 88 series.

Of the several series offered, it's the basic Firebird that is the cleanest. The TransAm with its front and rear spoilers is the best aerodynamically, while the mid-range Firebird Formula is the worst, with its forward-facing air scoops on the hood.

Pontiac's Grand Prix is not as bad as it looks. The grille is high, but it has a bow-shape, which helps the air flow. The headlamps do not protrude or make pockets, and the fact that the hood rises above the fender level is not a serious drawback. Its windshield is nicely wrapped around the corners, with a sharp rake, blending easily into the roof panel.

MGB, Spitfire and TR-7

The MGB has developed into a very neat aerodynamic package over the years. The new soft front, combining the bumper and air intake, with the grille totally eliminated to the chagrin of the traditionalists, presents a friendly aspect to the wind, giving a smooth transition into the hood area. Its roofline, in profile, is rather better than on the Porsche 911. Areas that could be further cleaned up are headlamp mountings and windshield corners, but that's not to say that the MG design shows glaring errors in these respects. They are not exactly wrong the way they are, merely perfectible.

The Triumph Spitfire has a well-curved nose and a smoothly sloping hood, and the windshield curvature is quite favorable. The car's worst aerodynamic drawback is the convertible top, which eliminates a smooth transition from roof to rear deck.

The TR7 on the other hand is a very modern design, with a grille below bumper level and concealed headlamps that protrude only when in use. Ridges and louvers in the hood were made to aid air flow and, at most speeds, do. The wedge-shaped profile and the dramatically cut-off rear end are somewhat questionable aids to streamlining, but certainly far better than a bulbous fastback shape would have been for this car.

Cordoba, Rolls, Mercedes and Beetle

On the Chrysler Cordoba, front fender edges protrude to make very sharp corners. The grille extends upward to full cowl height, with a plate-like face to the wind. Not quite as bad as the Rolls-Royce, with its knife-edge "greek-temple" grille, but certainly another example of how stylists' ideas can cause aerodynamic problems.

As early as 1927 Sir Henry Royce argued against the traditional radiator, on the grounds that it was aerodynamically bad design. But the business leaders of the company told him that the radiator had become a symbol, and

As if to repudiate aerodynamic laws, Rolls-Royce persists in its use of the squared-off vertical radiator. The entire front end is an aerodynamic disaster. The rest of the body is no worse than most sedans.

was a sacred element of recognition of the car as a Rolls-Royce.

The Silver Shadow still has the same grille, though it has grown lower and wider over the years. But the faults of the Rolls-Royce design do not end with the grille. They begin there, and go on for the whole length of the car. Front fenders are not well rounded, and carry excrescences such as screw-on parking lights, looking for all the world like you had mail-ordered them from J. C. Whitney and put them on yourself. The windshield has a very slight arc, with resultant sharp corners, and the glass stands unfashionably upright.

Mercedes-Benz is less religious than Rolls-Royce about its traditional honeycomb radiator, which has developed into a low, wide, and curved shell covering the front air intake. Still, it causes considerably higher drag than the horizontal air intake of the SL models, which also permits a lower and more sloping hood line.

Elsewhere in Mercedes-Benz bodies, you'll find that aerodynamics have been carefully studied and made use of to keep tail lights clean, to keep mud from the door sills, and to give other practical advantages. But aerodynamics have not affected the basic shape of Mercedes-Benz cars to any large extent.

As a final example, take the VW Beetle. It is such a curious melange of good and bad that one almost ends up admiring it for transgressing against the laws of aerodynamics with relative impunity. Indeeed, its drag coefficient is no higher

VW Beetle could be an object lesson in how not to streamline. Flat windshield and broken fender lines lead to vortex formation, while roofline stoops too abruptly to retain laminar flow.

than for many cars of far more recent origin. The blunt hood is an important asset, though it is probably not an ideal curve. In addition, the fenders are well rounded. That nose must be what saves it, for the rest is pretty hopeless, from separate fenders and a straight windshield to a fastback too steep to do much good.

chapter 5

fins, wings, spoilers and things

Sidewinds can drastically alter the lift forces acting on a moving car, as well as influencing the drag. Strong sidewinds and high-lift forces compromise the stability of the vehicle, and it's vitally important to the driver's ability to control his car that aerodynamic side forces and lift be kept within certain limits. A number of design elements play a part in reducing these forces, from the basic layout and weight distribution of the car, to the overall body shape, and the use of special anti-lift and stabilizing devices.

Before we discuss tailfins, wings and spoilers, it is necessary to take a few minutes to talk about sidewinds and their effect on a moving body. If there is no wind, and the car is traveling straight down the road, the air pressure distribution on the body surface will be symmetrical for the right and left sides of the car. Light or strong, the pressures are equal and opposite to each other, and therefore cancel each other out. As a result, there is no aerodynamic side force acting on the body.

Normal Wind Ruins the Laboratory Picture

This picture of pressure distribution changes as soon as a natural wind is encountered. A sidewind hitting the car at an angle relative to the car's direction of travel will apply pressure on the windward side of the body. How much pressure depends on the wind speed and direction, and, of course, the speed of the car.

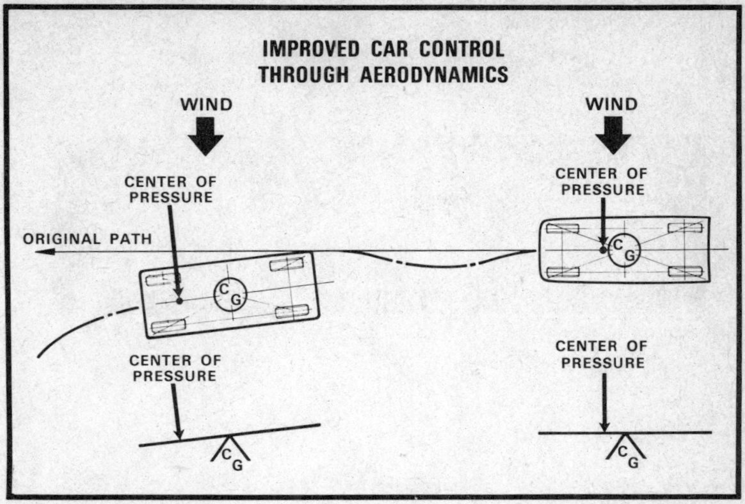

Sidewinds act through the car's center of pressure, just as centrifugal force acts through the car's center of gravity. Distance between the two is a measure of the car's aerodynamic stability.

Although the side force is actually applied over large areas of the body surface, it can be regarded as acting through a single point, in the same way that centrifugal force is applied to every part of the entire mass, but is regarded as acting through the car's center of gravity.

Aerodynamic forces do not act through the center of gravity but through the center of pressure. Like the center of gravity, it is an invisible point, an imaginary point. Both are important in terms of aerodynamic stability, for it is the distance between them that determines the car's resistance to sidewinds.

All passenger cars have their center of pressure located well ahead of the center of gravity. As a general rule, it is found in the cowl area in family sedans, and in the front wheel area for GT coupes. For ultra-streamlined prototypes it may be up front, near the headlights or radiator, and in extreme cases, completely ahead of the car. As a general rule, bodies with the lowest drag coefficient have the farthest-forward center of pressure.

Better Streamlining Worsens Stability

The car is also more sensitive to sidewinds the greater the distance between the center of pressure and the center of

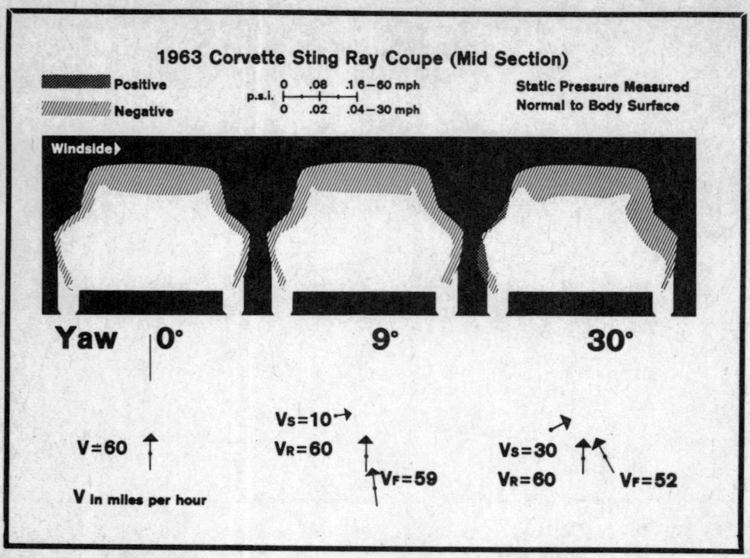

Pressure distribution on the body surface changes when there is a sidewind. With positive pressure on one side and negative pressure on the other, the car will tend to drift off course.

gravity. In other words, as you improve the car's slipperiness through the air, you compromise its stability.

The picture is further complicated by the fact that the center of pressure—like the center of gravity—is located at a higher level than the vehicle's roll axis. As a result, a sidewind can induce some body roll in the car, with the risk of increased instability which roll gives rise to in certain types of car. The old swing-axle type of VW Beetle had this problem.

But clever suspension engineers knew how to take advantage of wind-induced roll to develop a self-stabilizing vehicle response to sidewinds. Briefly, the commonly adopted method consists of providing increased roll understeer. How this is done is explained in another book published by *Tab Books, Chassis Tuning* (No. 2004).

A strong sidewind, acting through the center of pressure, will tend to swing the whole car around its center of gravity, and drive it off its intended course. If the wind is steady, the driver can counteract it by steering into the wind, and adjusting steering wheel position until the steering angle is balancing the side force. But sudden gusts are more difficult to cope with. It calls for frequent and very rapid driver reaction. The higher the speed, the greater the problem.

There is no way to provide an equal and opposite side force, acting through the same center of pressure. But there is a way to add a comparable side force in the same direction, acting at some point aft of the car's center of gravity. This would restore directional stability. Total sideforce would be about double, but the yawing moment would be eliminated. The body would feel sideways pressure, but at or near the center of gravity, with no effect on the car's heading.

The Effectiveness of Tail Fins

How can such a counterbalancing force be applied? It's an old aeronautical trick. It's called a tail fin. It works by providing a sort of sail for the sidewind at the opposite end of the body from the center of pressure, with the center of gravity acting as a fulcrum.

More and more, car designers have developed ways to use other tools in production cars, so that the need for tail fins is practically eliminated. Such tools include front wheel drive, for instance. Aerodynamic tools also exist, but are little used in production-car bodies, since sidewinds usually do not reach

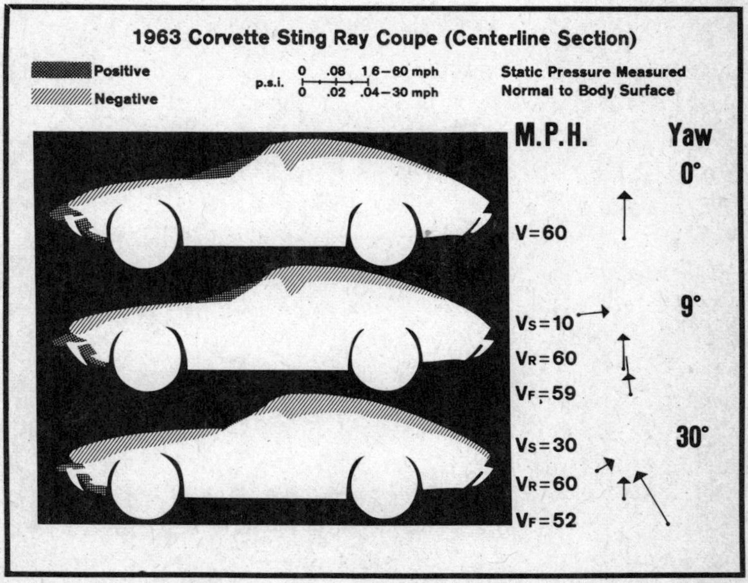

Sidewinds aggravate lift forces, as shown by this example. Positive pressure at the windshield base is almost eliminated at 30° yaw, altering the air flow for heating and ventilation.

critical strength within their speed range. For racing cars, it's another story.

The biggest tail fins are found on the fastest machines. That's such equipment as Breedlove's Spirit of America, Gabelich's Blue Flame, Frederick's Courage of Australia, and the Honda Hawk, built specifically to attack the world land speed record. They all follow the same basic design concept: A long cylinder with a pointed nose and a single giant tail fin.

Despite the high speeds attained by Formula One Grand Prix cars, they do not use fins. Why not? They have open wheels, open cockpit, wedge shape, and front and rear spoilers, which contribute to raise the drag coefficient. Consequently, they have a reasonably midships center of pressure.

Sports/racing prototypes also do without fins, in general, even if they have enclosed wheels and a closed cockpit. Again, other aerodynamic tools are used, such as the wedge shape, spoilers, etc. Even tire size plays a part—not in reducing the side force, but in resisting it.

Spoilers to Reduce Lift

A front spoiler is essentially a small panel attached below the nose, working as a deflector plate. It can be wide or narrow, deep or shallow, straight or curved. It can be positioned vertically, or projecting forwards at the lower lip. Its purpose is to partially block the air from passing under the car, and thereby reduce lift on the front part of the body. It does not add greatly to frontal area, and if its dimensions and angle are correctly chosen, it will not add to the overall drag.

Ford GT-40 prototype was very clean aerodynamically but several incidents occurred due to lift problems at high speed. They were cured by front and rear spoilers.

Ford Mark II at Le Mans in 1967. Very low rectangular duct scoops air for the low, tilted radiator, and the air escapes into the low-pressure area on the forward section of the hood. Rear spoiler reduces lift on the engine cover.

Do not confuse front spoilers with air dams. The air dam is a full-width panel, extending down to practically touch the roadway. It effectively blocks air from getting under the car from the front. Some air dams curve around the corners and extend back to the front wheels, so that the air cannot get under the car except behind the front wheels. Such air dams are used on modified production cars with great success. The air dam does add to the frontal area, but its drag-reducing potential is very high. Above all, it does, like the front spoiler, effectively reduce front-end lift.

An example will show how much difference a spoiler can make at high speed. Lift problems were the bane of the first Ford GT-40, built in 1964 to win at Le Mans. At 200 mph it had a total lift of 696 pounds—528 in front and 168 at the rear. Sidewinds would seriously aggravate the situation, and a 200 mph wind hitting the car at a 15° yaw angle would increase the lift to 1170 pounds (786 in front and 384 in the rear). Since the static weight on the front wheels was only 766.5 pounds (for a total of 1,825), the front wheels would then actually become airborne.

Experiments with a high nose were soon abandoned because it gave rise to even higher lift forces. A low nose, with cooling air extractors in the hood, was adopted. This shape gave front lift at 200 mph of 445 pounds, rear lift 422 pounds, for a total of 867 pounds. But in 15° yaw, lift on the front end still went to 704 pounds, which was too close to the static weight to allow safe running at such speeds.

At this point, Ford added a front spoiler, located below the air intake, which reduced front lift at 200 mph to 236 pounds, but raised rear lift to 272 pounds, with no sidewind. At 15° yaw, the front lift still did not exceed 309 pounds, or the rear lift 343 pounds.

Rear spoilers are added not below the tail, but on top of the deck, along the rear edge. It can be small wedge, giving a sort of ducktail profile, or a larger airfoil integrated with the body.

The basic purpose of the rear spoiler is to reduce rear end lift, but in many cases, especially with fastback designs, it has been found to also lower the drag. It will have no effect on pressure distribution on the front of the car, but it has proved capable of reducing lift along the whole length of the roof.

Using both a front air dam and a rear spoiler on the Ford Capri reduced total air drag by 11 per cent, and gave a 23 per cent drop in front-end lift, with a 73-per cent drop in rear-end lift. On the smaller Ford Escort, the same features reduced air drag by 14 per cent, and gave a 28 per cent cut in front end lift, with a 57 per cent cut in rear-end lift.

The Elegant NACA Duct Solution

A popular feature on modern racing cars, NACA ducts are not concerned with sidewinds and stability, nor with lift control, but strictly with reducing drag at the various air intakes needed on a car. Will they ever be used on production cars? It is doubtful that they would be of great value at normal highway speeds.

What are now known as NACA ducts stem from research work performed in the 1945-55 period by the National Advisory Committee for Aeronautics, the forerunner of the National Aeronautics and Space Agency. Turbojet aircraft, which were then in their experimental phase, needed air for various functions, just as planes powered by reciprocating engines had needed air. But the jet planes were much faster, and it became much more important to make the ducts efficient on jet planes.

The NACA research provided certain immediate conclusions that are still valid, and apply equally to high-speed ground vehicles. Nose inlets are efficient, but sometimes require long and sinuous internal ducting, which means increased drag. In addition, they need grille protection to avoid taking in debris. Air scoops can be effective only if they extend into the air flow beyond the boundary layer, and if placed close to where the air is to be used.

Submerged inlets are naturally protected against debris and involve lower external drag than air scoops. But their location is very critical. They can be effective only if positioned in a high-pressure area and provided with the proper entrance geometry. It is the submerged type of duct that has been named NACA duct in common parlance.

Their design has become a very precise specialty of great importance for racing cars. A few of the design parameters are worth mentioning. First, there is the duct entrance aspect ratio, which means the ratio between the width and depth of the opening. It has been established that the best results are obtained when the inlet is three to five times wider than it is deep. The ramp angle, which defines the difference between the external sheet metal and the duct floor, should be kept within a specific, narrow band of about 6 to 11 degrees. Too large a ramp angle will lead to boundary layer separation on the sides of the duct and possible spillover from the external air flow. Too small a ramp angle will lead to higher drag losses as a result of boundary layer buildup inside the duct. Invariably, a flat floor is preferable to a curved one.

The walls give best results if they diverge, i.e., spread out from the front. In other words, the ramp becomes wide as it deviates from the outer skin, and reaches its widest point at the lip, which forms a door lintel or eave over the inlet. The walls need not be straight—many designers have obtained favorable results with a champagne-bottle plan form.

Wings Reduce Lift, Increase Drag

From front and rear spoilers to free-standing wings was a big step structurally, but a small step in logic. The "wings" are negative-lift airfoils carried on tall struts extending beyond the boundary layer.

Wings were first used in Formula One racing at the Spanish Grand Prix in 1968. Brabham and Ferrari cars used small air-foils mounted on slim pillars, both back and front. They were fixed in position, and had very little effect until very high speeds were reached, when they substantially reduced the lift. Then Lotus developed a big, wide, adjustable wing, which was first used in the French Grand Prix at Rouen in 1968. After that, all the Grand Prix and Can-Am car constructors developed variable-geometry wings.

Despite the ability to tilt the wings, some cars suffered a surprising drop in top speed, which was, of course, due to an

Wing to produce negative lift is placed immediately above the rear wheels on the 1971 Can-Am McLaren M8E racing car. Cooling air is extracted through louvers in fender tops and into wheelwell air flow.

undesirable side effect from the wings: higher drag. Lift drag is the same whether the lift is positive or negative. The greater the high-speed stability benefits from the wing, the lesser the need for this effect, due to the lower top speed. On the Chaparral, wing angle was controlled by a pedal operated by the driver's left foot. In case of a failure in the mechanism, the wing was arranged to return to the position of maximum down-thrust (fail-safe).

Methods for varying the wing angle differed for the various makes of car. On the 1968 Matra V-12, the wing position was controlled by a lever directed by a worm-and-nut arrangement in which the worm was driven by a small electric motor. The system worked automatically, being activated in accordance with brake application and gearshifts. On the Nissan 381 which won the Grand Prix of Japan in 1968, the wing was split in two, each half being separate controlled in accordance with suspension movement, acceleration and braking.

The early racing-car wings were attached to the body but then Jim Hall found a better way, which was soon adopted by the majority of constructors. Hall's invention was the Chaparral wing mounting, with struts extending directly from the wheel hub carriers, avoiding any connection with the body. Thus, the suspension system was freed from carrying the negative-lift load, which was placed directly on the road wheels where it could do the most good.

After a number of accidents (caused by wing malfunction or disintegration), the wings were outlawed, except when integrated with the bodywork and coming within specific dimensions.

The case against the wings rested on two main points. First, the wing was subject to enormous aerodynamic forces, which meant that robust supports would be needed to withstand them successfully. Since robust construction means more weight, and the engineers are always trying to eliminate extra weight, the results were a lot of flimsily constructed wing struts and stays. How great are these aerodynamic forces? The wings on the 1968 Lotus produced 200 pounds negative lift on the front end and 400 pounds negative lift at the rear end at a speed of 150 mph.

Secondly, there was the matter of car-to-car interference through the turbulence created by the wings of cars passing each other or running in close formation, which could jeopardize the stability and controllability of both machines.

The Unsatisfactory Wedge Body Solution

In their quest for negative lift, racing car constructors were led to the idea that the body itself could be given an airfoil silhouette. The wedge body was born. Apparently, the designers believed that pushing a wedge through the air, with the thin end in front, would bring considerable air pressure on the entire upper surface. Well, it was partly true. But it was

High wing on 1966 Chaparral was mounted on struts extending directly from rear wheel hub carriers, leaving the body unaffected by the aerodynamic forces acting on the wing.

Wedge shape is typified in this Pininfarina-bodied Alfa Romeo Tipo 33 from 1971. The wedge was intended to produce positive pressure on the entire upper surface of the body. It was not very effective.

soon discovered that wedge-shaped bodies were inordinately sensitive to *attitude*. By that I mean the wedge angle. If the static base line is parallel to the ground, it will not take more than a few degrees of pitch movement to change negative lift into positive lift.

Ford made some tests which showed that a typical wedge-shaped body would generate 100 pounds negative lift at a given speed, running in its horizontal attitude. Raising the thin end to put the wedge in a 4° nose-up attitude would produce about 600 pounds of lift!

Because pitch is unavoidable under actual driving conditions, it became necessary to provide the wedge with a static nose-down attitude. The deviation from horizontal would depend on suspension parameters such as maximum spring deflection, and aerodynamic design elements influencing the body's pitch moment. Most designers have given up on the wedge shape, mainly because it created more problems than it solved.

Porsche 917

One car combined so many stabiilizing and anti-lift devices that it deserves special treatment. That's the Porsche 917, which won every race Porsche intended to win, and retired undefeated. The basic shape came from Porsche and was mainly the work of Helmuth Bott. It was developed in the wind tunnel of the Motor Vehicle Research Institute in Stuttgart, and was first shown in March, 1969 in its Langheck (long tail) version, with a C_D of 0.33. The front fenders carried anti-lift deflectors, shaped like stub wings, and fixed at about 20° from horizontal. On the sloping tail were small fins carrying a crossbar at their rear end. This crossbar served as a mounting for two flaps extending backward, about 15 inches wide and 12 inches long. Their angles were individually controlled via a mechanical linkage to the rear suspension, so as to raise the rear edge during rebound deflection and return to near-horizontal in jounce.

A Kurzheck (short tail) version appeared in June. The body was simply cut off at about two o'clock, looking at the rear wheel of a car headed to the left. The flaps were retained, but in new shape: about 25 inches wide and 3 inches long. Their range went from about 20° to 90° (vertical). The 917 K had a C_D of 0.40.

For 1970 the Langheck went to wedge shape. The rear fenders rose like tail fins with a 20-inch rectangular cross-section, leaving an open canal between them. Fixed spoilers were added to both fender tips. NACA ducts admitted air for tire cooling. At the front end, the stub wings disappeared, and the air intakes were redesigned.

This version had a C_D of 0.45, and went up to about 0.46 for the Le Mans version, on which the canal between the rear fenders was fitted with an airfoil to provide negative lift needed for the very high speeds reached there. At the same time, NACA ducts on both sides of the engine cover were added.

For 1971, the 917-L reverted to a rear fender line resembling the original smoothly sloping profile. The fender skirts were extended even further, and tall fins rose from both, carrying a high wing between them. Wing angle was controlled by rear suspension movement, as with the flaps used on earlier

Turbocharged 1000-hp Porsche 917 sacrificed streamlining to anti-lift devices. Air-dam nose almost touches the ground. Huge rear wing was fixed between tail fins, but its aileron flap was controlled by suspension movement.

versions. A redesigned wedge-type nose was adopted. This was the fastest Porsche ever built. With its C_D of 0.36, it was timed at 236 mph down the Mulsanne straight at Le Mans. Team Martini's race-winning 917 L was a new cross-breed, however. It had the wedge-type rear fenders, no flaps, but high tail fins. The 1971 version of the 917-K also adopted tail fins, and some had open cockpits.

The ultimate in the series was the 917-10 developed for the 1972 Can-Am races. With 900 to 1000 hp on tap, drag was less important than lift, and the whole shape was a masterly combination of anti-lift devices. But the car's C_D was high: 0.65! It had an open cockpit with dual headrest-cum-rollbars faired into the engine cover. The nose had a horizontal lip to get a stagnation line as close to ground level as possible. The air intake was a scoop blending into the NACA-ducted hood, while the front fenders had sort of an anthill profile, with air-extraction louvers from 10 o'clock to noon. Rear fenders had just a hint of wedge in their shape, but they carried enormous tail fins, rising smoothly from ahead of the wheels and extended banner-like backwards from the sharply cut-off fenders. Between them was a high fixed wing, with an adjustable aileron at its rear edge.

chapter 6
history of car streamlining

Early awareness of the problems of automotive aerodynamics contrasts strangely with the slow adoption of streamlining by the auto industry. Back in 1911, W. G. Aston wrote in the *Autocar*: "During Autumn, when days are calm and leaves litter the ground in some quantity, the way in which air is carried along by a car is very noticeable. Light objects fifteen or twenty feet away from and at the side of the car are affected by its passage, and this gives a good idea of the very large volume of air which is uselessly set in motion."

Most cars of that time were designed without regard for the disturbance their motion caused in the air around them. Coach-work design was largely inherited from the makers of horse-drawn vehicles. Quality of materials and workmanship counted for much more than the body shape, and its aerodynamic efficiency was totally overlooked.

And yet, by 1911, many countries had a nascent if not flourishing aviation industry, and the "teardrop" shape of the Zeppelin dirigible balloons was a familiar sight. The reasons for its shape were also well known. It was to be many years, however, before scientists would realize that a teardrop in free fall does not at all resemble a Zeppelin or any other idealized geometrical shape—it consists of one large spherical drop with a number of smaller spherical droplets in its wake.

By 1911, in fact, a large number of racing cars had been built, equipped with bodies inspired by the Zeppelin. Quoting a shrewd, old-fashioned engineer, W. G. Aston described the

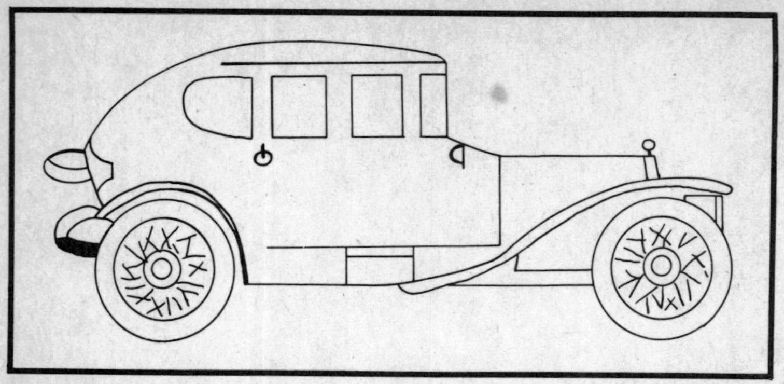

1910 Gregoire featured C-curved windshield and an egg-shaped tail with an extended spare wheel cover. But no streamlining was done on the nose.

racing cars as being "as far as he could see, employed as a means for ascertaining what is to be the future shape and form for the envelopes of dirigible balloons."

It was in racing that speed was most important, and the limits of power most felt. Consequently it was not surprising that streamlining should first come into use in some form of racing. Logically, it came first for speed-record-setting machines, and later for road-racing and track-racing.

The Early Streamliners

The first to do something dramatic about streamlining a car was a Belgian racing driver named Camille Jenatzy. He was a proponent of electric cars and decided to make one that would set a new speed record. For that, even in 1899, he knew

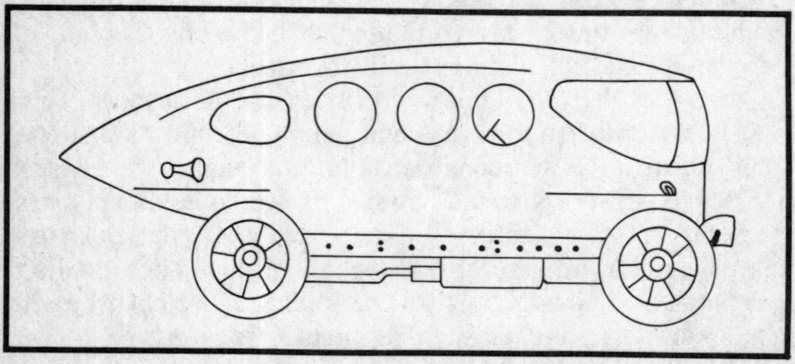

Count Ricotti designed this car in 1913. The chassis remained standard, and neither driver's seat nor steering wheel was moved forward to take advantage of the shape.

streamlining was vital. He designed a bullet-shaped body, of cylindrical cross-section, with tapered nose and tail. It was built by the Maison Rothschild. But Jenatzy could not reconcile the bullet shape with the need for wheels, so he just gave up, and placed the body on top of the tall chassis, and left the four wheels out in the open air. However, his car was fast enough to do the job, covering the flying kilometer at a mean speed of 65.8 mph (one way only, and with the wind behind him).

Another strange electric car appeared in 1902. Built by Walter Baker, it was a low-profile cigar-shaped vehicle with a closed cockpit. It was wrecked on its first attempt at the world land speed record, due to a tire blowout, but it was the first American streamliner, and therefore historically significant.

This was followed by a fantastic exploit in 1906, when Fred Marriott drove a Stanley Steamer with a low-profile boat-shaped body on the beach at Daytona, covering the flying mile in 28.2 seconds for a mean speed of 127.66 mph. This record was vouched for by the Automobile Association of America, but it was not ratified by the Association Internationale des Automobile Clubs Reconnus (forerunner of the Federation Internationale de l'Automobile).

Rumpler Tropfenwagen (drop car) created a sensation in 1921. But Rumpler's concept was largely two-dimensional, as the roof line was allowed to maintain its full height practically throughout.

Sterkenburg of 1933 had radical sloping hood and neatly tapered roofline. This car became the basis for the Lincoln Zephyr, which used its broad concept but conceded much to contemporary fashion.

About 1910 designers tentatively began to try out their streamlining ideas on touring cars and family cars. One note worthy example is the Porsche-designed 1910 Austro-Daimler "Torpedo" built for the Prinz Heinrich trials—an open body with smooth-paneled sides. It won the event, but more important, it was to influence touring-car design for a decade.

Attempts to streamline closed cars produced results that most of us today would call weird. A special body by Alin & Liotard on a 1910 Gregoire showed several important streamlining elements. It was a three-door sedan with a balloon-shaped rear end, finished off by the horizontal spare wheel cover. A narrow, vertical windshield was flanked by two small pieces of curved glass, rounding off the corners, and creating a form of panorama-windshield. The front end, however, did not depart from normal practice, with free-standing headlamps, vertical radiator, and wind-catching fenders.

Count Marco Ricotti was an engineer who was involved with both cars and airplanes, and it was obvious to him that the car designers could learn much from looking at flying machines. In 1913 he designed a teardrop-shaped car body, and ordered it built by Carrozzeria Castagna. The chassis was a standard Alfa with a 50-hp engine, for a touring car which in regular production form had a top speed of about 55 mph. With the Ricotti body, its top speed was 87 mph.

The First World War Accelerates Development

Aircraft design and the study of aerodynamics matured fast under the pressure of war, and after 1918 some of the lessons learned began to show up in automobiles. The first car conceived as a total streamlined package was the 1921 Rumpler prototype. It was designed by Edmund Rumpler, a Viennese engineer who had gone to Berlin as a consulting engineer but ended up as an aircraft manufacturer, producing large numbers of the Taube monoplane during World War 1. His car borrowed ideas from both aircraft and boats, however.

The "teardrop" plan view was achieved by placing the engine in the rear and moving the windshield and driver up front. The windshield was curved, and the fenders were small horizontal panels of minimal frontal area. But the overall impression is still somewhat two-dimensional, for the body retained nearly full height throughout its length. Rumpler had not completely grasped the spatial nature of aerodynamics.

The first of the real scientists was Paul Jaray. In 1914 he became chief of design and development for the Zeppelin airship works, and in 1920, with Germany's aircraft industry outlawed, he began to study car aerodynamics.

The first Jaray-designed body was built in 1922 on a Ley chassis. His patented formula was to design the entire body up to the waistline as an airship half-body, and then add a smaller half-body for roof and windows.

In 1923 he set up his own business in Switzerland, and sold licenses for his streamlined car body to many auto makers, over the years to include BMW, Fiat, Adler, Auto Union, Hanomag, Maybach, Mercedes-Benz and Opel.

Chrysler designed the Airflow without his help, but it infringed Jaray patents, and so Chrysler paid royalties to Jaray for each streamlined Chrysler and De Soto built. Tatra bought a Jaray license to be clear of infringement, but designed its Type 77 of 1934 without actually consulting Jaray or drawing know-how from his patents.

Jaray was the first to make wind-tunnel tests on scale-model cars, starting in 1922. Wind tunnels had been around longer than anyone knows. According to the American Heritage History of Flight, the first wind tunnel for aviation research was built in 1871 by two Englishmen, F. H. Wenham and John Browning. Of course, it was used primarily to test theory, rather than actual planes. Research Trends, a

True Jaray streamlining at its best is found in this 1938 Adler. The front fenders blend into the body sides, and the roofline has a long, gentle taper.

periodical published by Calspan (formerly Cornell Aeronautical Laboratory) points out that a wind tunnel, 16 inches square, and six feet long, existed in Dayton, Ohio, at the turn of the century. Two years before the Wright brothers flew at Kitty Hawk, they were comparing lifting surfaces in this tunnel. With regard to visualization of air flow, a French pioneer, Etienne-Jules Marey, was using smoke streams in a tunnel as early as 1895.

Jaray's half-body approach led racing-car builders to develop "tank" bodies (without the upper half-body). In 1924-25 both Bugatti and Chenard-Walcker in France built and raced such cars. Others focused their attention on the family car.

Emile Claveau, who later founded the Bureaux d'Etudes et Recherches Techniques in Paris, built a streamliner with forward control in 1927. It was a design with a tapered box-section fuselage and open wheels—a big improvement on the Rumpler, but still ignoring many of the problems.

Land Speed Record Inspirations

About this time, the world land speed record stood at over 200 mph, and high-speed stability problems had to be faced. Malcolm Campbell's Bluebird of 1928 was first with fins—two vertical stabilizers attached to the body inboard of the rear wheels, above the fairings for the axle.

Lift problems had also been experienced, and scientists quickly found solutions. Fritz von Opel's rocket-propelled speed-record car, Rak-2, of 1928, used negative-lift airfoils, extending like aircraft wings from the cigar-shaped fuselage.

But what about mass-produced cars? Styling was a domain restricted to specialist coachbuilders (many auto makers were still producing chassis only in 1927). General Motors was about to set up its own "Art & Color" section, predecessor of what is now GM Styling. Ford was phasing out the Model T, but the Model A was still a box on wheels. And no mass-producer showed any interest in streamlining. Yet progress was imminent.

Ford began using a wind tunnel for test purposes as early as 1929, when the company was developing the Ford Trimotor airplane. The test section has a circular cross section of four feet, with an air speed range up to 110 mph. In 1930, Ford began to use this tunnel for new car shapes as well as for aircraft work. The car models were built to one-quarter scale. It was in 1936 that Ford completed its own full-scale wind tunnel, with a test section 9 feet high and 11 feet wide. Air speeds up to 75 mph were possible in this tunnel.

Finally, Attention by Mass Producers—and Kamm

The first tunnel designed specifically for full-scale research on automobiles was built in 1930 at the Motor Vehicle Research Institute in Stuttgart, Germany. This establishment was directed by a Swiss-born engineer named Wunibald Kamm, who had formerly worked six years for Mercedes and four years for the German Aviation Test Institute in Berlin. With his new wind tunnel, he began a series of experiments which resulted in epoch-making discoveries. He was the first to enunciate firm theories of boundary layer behavior, and gets credit for having brought car aerodynamics out of the empirical field and into the realm of science.

Tatra 77 was an original creation by Hans Ledwinka, but Tatra was also licensed to make use of the Jaray patents. Protruding headlamps conflict strangely with the "soft" hoodline. The small, single central tailfin was found not to be effective.

Kamm-inspired Superleggera body on a 1940 competition-type BMW 328 shows how the designers at Carrozzeria Touring missed the real principle of the K-heck, while adopting many Kamm design elements.

In 1935, he created the K-body which still inspires designers the world over. Most of the K-body prototypes were built on BMW chassis. The main feature of the K-body is the rear-end, which the Germans call "Heck." Kamm's "Heck" had a mysteriously high roofline, which ended abruptly, almost like a station wagon, with a small step at the belt line. It was a milestone in lowering the drag coefficient.

In technical terms, the Kamm body can be described as the car with the smallest interrupted cross-section. In plain language, it means that the car isn't going to be much lower or narrower than at its widest and tallest anywhere behind that point, in the interest of developing the most advantageous flow lines for the air around it.

In other words, the roofline could only taper a few degrees without boundary layer separation occurring. Kamm reasoned that the cross-sectional area affected by turbulence is smaller when the body tapers very slightly right to the end, rather than when fastback rooflines of more severe taper are used. In many instances, cars that looked very streamlined would suffer boundary layer separation on the roof at the mid-wheelbase point, while a typical K-body would retain laminar flow beyond the rear wheel center line. As the points on the roofline where separation occurs can differ considerably in height, it's easy to see that the K-body actually brings laminar flow to a lower level than the typical fastback—and that's why it has lower drag.

Kamm's work was not known in Detroit when Chrysler came out with the Airflow in 1934. Its basic form was final as early as 1932, after about two years of testing. Chrysler had its own scale-model wind tunnel in operation as early as 1930. The design work is credited to Carl Breer, a member of the famed engineering triumvirate of Zeder, Breer, and Skelton, creators of the first Chrysler car and leaders of Chrysler engineering until the 1950s.

The Airflow did have lower drag than the regular Chrysler, mainly due to its short, blunt hood, enclosed headlamps, with lesser benefits from its split-V windshield and fastback roofline. Still, Breer was disappointed that the real car failed to equal the performance predicted in scale-model tests. Few buyers wanted the Airflow, and it was discontinued in 1937.

Another significant attempt at streamlining an American car was the Sterkenburg prototype, built by Briggs Manufacturing Co. in 1933 to designs by John Tjaarda. It was perhaps a little wild, but a mixture of fashion and function that appealed to Edsel Ford. Ford took over the design, and used it as the basis for the first Lincoln-Zephyr of 1936.

The Rash of Experiments in the Thirties

In 1933, Pierce-Arrow built five prototypes called Silver Arrow, designed by Philip Wright, with special fastback bodies incroporating fender lines that blended into the body sides. Headlights were faired in. But the car had a tall if sloping radiator grille, and a flat-glass nearly upright windshield. A year later, Hupmobile supplemented its regular lines with a "streamliner" designed by Raymond Loewy—but this was not a true aerodynamic study, and very few were produced.

Chrysler Airflow was a breakthrough, but it was killed by the popular vote. People bought cars with long hoods and upright grilles. Aerodynamically flawed in many ways, the Airflow was nonetheless a milestone.

Cord of 1936 was dubbed "coffin-nose" for its unusual styling. Not designed as a streamlining exercise, it nevertheless was a big improvement on the typical American car of its time.

The 1936 Cord, created by Gordon Buehrig, was a masterpiece of industrial design, observing certain rules of streamlining while ignoring others. At the same time, independent inventors and designers all over America were busy with their own interpretations of streamlining, and created a variety of attention-getting prototypes, though their scientific value remains questionable.

Probably the first was Captain James V. Martin, who produced a streamlined prototype with a rear engine in 1931. The most scientific was probably the Stout Scarab of 1935, a long-wheel-base streamliner with sloping, short hood, and a bulbous tail housing a rear engine, designed by William B. Stout, an outstanding engineer with high credits in both automobiles and aviation. The weirdest was R. Buckminster Fuller's Dymaxion of 1933, but it should not be judged too severely for its shape, since the designer was, above all, experimenting with new construction methods and materials, and less concerned with aerodynamics.

The well-known biologist, Calvin B. Bridges, who lectured at the California Institute of Technology, became interested in car aerodynamics, and in 1936 built a streamlined three-wheeler with full-width body.

Other privately built streamliners were those of Benjamin F. Harris in Champaign, Illinois (a front-wheel-drive V8 of 1937); Ralph Lysell's Auto-Avion of 1937, from Philipsburg, Pa.; Professor Moddie's 1934 tricycle with enclosed wheels;

the Heinz of 1938 (another front wheel drive prototype); Harry S. Parker's Terraplane-based model built in Ellsworth, Maine, in 1936; the Pribil Aircar built in 1936 by the Saginaw Stamping & Tool Co.; and Wellington E. Miller's 1936 Advance V8 three-wheelers, built by the Advance Auto Body Works in Los Angeles.

While these experimenters were in the limelight, the real scientists were making quiet progress behind the scenes. One was Professor W. E. Lay, who lectured at the University of Michigan in Ann Arbor. Starting in 1933, he conducted a series of tests on the effects of small variations on simple basic shapes. Lay developed novel designs and new theories that were to have considerable influence on further research work.

Elliott G. Reid was a professor at Stanford University in California, who supervised the wind tunnel tests made by two of his graduate students on 1/5-scale car models in 1933, and wrote some excellent reports on automotive aerodynamics. Reid and his students also developed a low-drag car body shape, not unlike a smoothed-over Claveau with faired-in wheels.

R. H. Heald was an official of the National Bureau of Standards who became interested in aerodynamics research in the early 1930s. He ran an important series of wind tunnel tests on sidewinds, stability, and tail fins in 1937, and laid down some of the basic rules in this area.

Continuing Scientific Research in Europe

Behind Jaray and Kamm, other European scientists were also performing valuable research work in car aerodynamics. Ludwig Prandtl had developed a simple form of recirculation-type wind tunnel, with an open test section, at the University of Goettingen in Germany. His work led to the creation of the Goettinger Test Establishment for Aerodynamics, which in the 1930s was directed by Professor Seifferth. Together with his pupils, graduate engineers Schloer and Hansen, he created the body shape now known as the Goettinger form. It looks like half an egg, split along the major axis, on wheels. A full-size test car was built on a rear-engined Mercedes-Benz chassis in 1938.

Jean Andreau was a most versatile engineer who had attended the Saint-Cyr military college and became attached to the French army's technical service, where he made pioneering studies in artillery and aviation during World War 1. After the war he set up a test establishment for chassis and suspension studies near Bordeaux. He became a consultant to Citroen and Donnet, but in 1932 joined the Laboratoires Chausson and began scale-model wind tunnel tests. The bodies for the Peugeot 402 (1936) and 202 (1939) were based on shapes developed in Andreau's experiments. The long-wheelbase 402 broke new ground in drag reduction for family cars, with a C_D of 0.383, at a time when few production models had got below the 0.5 mark. In 1938, Andreau desgined a far more extreme body on a 402 chassis, with a single tailfin, which proved to have a C_D of only 0.28. Andreau also designed streamlined bodies for Delage racing models, and was a consultant to Captain George Eyston for the body shape of his 1938 Thunderbolt, which briefly held the world land speed record.

Jules Chausson, in addition to being chairman of his own company, was also a director of Chenard-Walcker, which may have provided the latter firm with an incentive to venture into streamlining. A spectacular prototype, "Mistral," appeared in 1933, but it's not known how much of its shape is due to Andreau, and how much was designed by Henri Toutee, the Chenard-Walcker engineer who had been responsible for the "tank"-bodied cars of 1925-29.

Emile Claveau never stopped developing new shapes, and in 1931 proposed a range of small cars with front wheel drive and tank-like bodies. The hood was short and well-rounded, and the fenders were semi-circular lips extending only about an inch or so from the slab-sided body, which contained the wheel housings. The designs show a notchback sedan, a coupe and convertible. However, the upper bodies and rooflines showed rather contemporary lines.

A remarkably advanced prototype was built in 1936 by Andre Dubonnet (scion of the aperitif family). His design used forward-control, with a tapered and sloping nose section, and the fuselage tapered off to the rear, with the rear wheels carried like outriggers, the wheels and their suspension completely faired in. The engine was placed just ahead of the rear wheel axis, and it was actually a Ford V8. The whole nose section opened as a door, giving access to the front seat. Side doors on both sides served the rear seat passengers. A gigantic tail fin surmounted the sloping roofline. Its streamlining was perhaps intuitive rather than scientific, but it gave highly convincing results. At the time when the normal-bodied Ford

V8 could only reach 81.4 mph, the Dubonnet prototype was timed at 108.35 mph.

Manufacturers' Serious and Styling Efforts

Throughout the 1930s there was a lot of pseudo-streamlining activity on the part of auto manufacturers, coachbuilding companies, and private inventors. Some were honestly trying to create wind-cheating shapes, while others simply engaged in new styling exercises. In the first category, we must consider the work of Sir Dennistoun Burney, who built the rear-engined Burney Streamliner in 1929 and sold the design to the Crossley Brothers in Manchester, who produced their own version of it in 1934-35. A commander in the Royal Navy, Burney had invented the mine-sweeping paravane in World War I and in the 1924-28 period he was the manager for the team that designed the British Airship R.100. His car had a long, tapered body of rectangular cross-section. This aircraft-type fuselage was preceded by a boat-like nose with recessed headlamps, but the front wheels were out in the open, with cycle-tape fenders. The Crossley version reverted to a traditional hood and radiator, with free-standing headlamps.

It's in the other category that you'll find Hermann Spohn of Ravensburg in Germany. His chief stylist, Josef Eiwanger, created a radical streamliner on a Maybach V-12 chassis in 1931, a car that may have been the inspiration for the Pierce Arrow. It probably also inspired Volvo to build a streamlined prototype in 1935, which led to limited production of the very modern-looking Carioca in 1936. Other Spohn bodies were regressive in many ways. On a 1935 Spohn-bodied Mercedes-Benz 500 Eiwanger left the entire front end back to the cowl untouched by streamlining, but added a sleek fastback coupe. However, in 1937 Spohn built a streamlined Maybach under Jaray patents.

Tatra Led Serious Production Streamlining

By that time, this design was no longer in the vanguard of streamlining's march, and the Jaray-Spohn Maybach was probably surpassed in terms of drag reduction by an earlier production model from Czechoslovakia: the Tatra 77. The prototype was made in 1933, and had a flat-face windshield and flush-fitting headlamps in the rounded hood. On the production model of 1934, the windshield was split in three, with small flat pieces to round off the corners. Little pods for the headlights emerged from the hood, probably to facilitate adjustment. The 77a appeared with an embryonic tailfin, as the car had exhibited serious stability problems at high speed. But in 1936 the 77a was replaced by the 87, on which the front fenders presented a roiunded front rather than the truck-like fender panels of the 77-77a. The air ducts to the rear-mounted V8 were considerably modified, and the 87 remained in production (with interruptions) up to 1948.

A number of less advanced cars gradually came into production, such as the 1939 Adler 2.5 and Hanomag 1.3. Volkswagen prototypes were streamlined to the extent their length would allow, as were some small cars produced by Steyr and Skoda from 1937 to 1940.

Fiat flattened out and rounded off the nose of its 1500 model in 1935, and adopted similar styling for its 508C in 1937. Even the tiny 500 had a rounded nose. But Fiat's first serious attempt at streamlining was a special-bodied Balilla for the 1939 Mille Miglia, which developed into a production model, the 1100-S, in 1947.

Panhard-Levassor had introduced vertical windshields with curved-glass corner pieces in 1933, more in the interest of improving visibility than aerodynamics. However, in 1937 the company began production of the Dynamic, which had a very streamlined body designed by Louis Boinier, with a sloping 3-piece windshield, recessed headlights, and fenders with partly enclosed wheels both front and rear.

It is fitting to end our review of the pre-World War II period with the Panhard-Levassor, for this company was also among the first to explore streamlining further after the war. In 1948 Bionier designed a futuristic body for the Dyna chassis, drawing heavily on modern aircraft shape for inspiration, and performing aerodynamic tests at the Institut Aerotechnique de Saint-Cyr as well as in the Laboratoires Eiffel in Paris. The result was the Dynavia, which reached a top speed of 88 mph with a two-cylinder engine. When crusing at a steady 62 mph it gave a gasoline mileage of 42 miles to the gallon! From this Dynavia prototype, Bionier developed the shape for the Dyna-54.

The Panhad chassis was also used for the D.B. and Monopole sports/racing cars whose top speed astonished the constructors of more powerful machinery when these little streamliners ran at Le Mans, and invariably won the Index of Performance and the Index of Energy.

During the war Jean Andreau had been toying with ideas for streamlining very small cars, and he sold his proposed VL-333 minicar to Mathis. It was a 3-wheeler with a single rear wheel. The body had a egg-shaped tail, with an elongated nose that neatly enveloped the wheels. He also designed a new family car, which was named Mathis 666. Both prototypes were exhibited at the Paris auto show in 1947, but neither got into production.

Emile Claveau had also spent the war years dreaming up new shapes, and showed his Descartes model in 1947. It was a slabsided four-door sedan, with nearly symmetrical hood and deck lines, presenting a very "soft" front. But he failed to get production under way.

Meanwhile, up in Sweden, the Saab aircraft company had decided to get into the car business, and Sixten Sason designed a small sedan with excellent aerodynamic properties. The prototype was shown in 1947, and production started two years later. It was a short-hood design with a very sloping windshield and a fastback roofline. It also had a completely smooth bellypan, and the fenders were shaped to minimize the drag set up by the revolving wheels.

Frustration in America, and a New Start

The trend toward streamlining of production cars in America received a setback when Chrysler killed the Airflow. The one-off Chrysler Newport and Thunderbolt, designed by Alex Tremulis and built in 1940, were convertibles, and only pseudo-streamliners. In my opinion, streamlining in the pre-war era culminated with the Packard Clipper, designed by Werner Gubitz.

After the war, a new start was made with the slab-sided Kaiser and Frazer bodies, designed by Howard Darrin. Another noteworthy departure was Frank Spring's 1948 Hudson, with its step-down frame and very stretched-out fastback roofline. The 1949 Ford, designed by Holden (Bob) Koto, was modern rather than streamlined, but the contemporary Nash, for which Pininfarina had acted as a design consultant, introduced new ideas in streamlining the American family car. A few years later came the Studebaker Starliner (1953), which was mainly designed by Robert

General Motors Firebird 1 experimental turbine car was almost a throwback to the 1906 Stanley Steamer, except for negative-lift ailerons and a tall rudder.

Bourke. Its low build alone made it outstanding, and the nose shape, with the sloping hood and very low split grille, gave it exceptionally low drag. Many years later, this nose shape reappeared on the Pontiac Grand-Am, and the sloping hood is finally becoming a feature of most new bodies from General Motors. But General Motors was a late-comer to streamlining having left the matter of shape entirely to the styling studios, unencumbered by obligations to face and circumvent the realities of aerodynamic effects.

It was not until 1951 that GM began tunnel testing scale models, starting with the experimental gas-turbine driven Firebird 1, rather than a production-car project. It was not GM's wind tunnel, but a facility set up in 1926 (by a Daniel Guggenheim grant) at the California Institute of Technology. For many years to follow, this was the extent of GM's aerodynamic research. Full-scale testing was not undertaken until 1969, when GM arranged to use the wind tunnel of the Lockheed-Georgia Company, which was designed primarily for aircraft.

To this date, the only wind tunnel at GM is an environmental wind tunnel, built in 1955 by Harrison Radiator Division to test engine cooling, heating, and air-conditioning systems. It has been used for some scale-model airflow testing, but it has severe limitations for such purposes.

It was about 1950 that Virgil Exner, chief stylist at Chrysler, became interested in streamlining, and made an arrangement with Carrozzeria Ghia in Torino, Italy, to build a number of prototypes. The first was the Chrysler K-310, which served as the basis for a paper on aerodynamics and design that Exner delivered to the Society of Automotive Engineers in 1953. The K-310 led to a number of new creations: De Soto Adventurer in 1953, Plymouth Explorer in 1954, Dodge Firearrow in 1954, Chrysler Falcon and Flight Sweep in 1955, and the Norseman (which went down with the Andrea Doria) in 1956.

The Ghia-Gilda of 1955 was a truly streamlined body, with a sloping hood, wedge-shaped profile, and tailfins. The Dart of 1957 went even further, and had enormous fins. Exner's further creations for Chrysler degenerated from true streamlining into the same kind of gimmickry that GM Styling had inflicted on its 1956 generation of "dream cars."

Italian Stylists and Scientists

Italian coachbuilders participated in the advance of streamlining throughout the 1930s and 1940s in several ways, but their creations were mainly intuitive or pseudo-aerodynamic. Their most scientific work before the war went into motorcycles, and later, into speed-record machines. Piero Taruffi, who had been associated with the development of both Rondine and Gilera motorcycles in the 1930s, created a twin-boom car in 1950. It was like two side-by-side torpedoes, one containing the engine and the other the driver. Enrico Nardi built a sports/racing car to the same formula in 1955, but closed in various inter-boom connections.

Alfa Romeo had no background in real aerodynamic research when the company decided to build an experimental sports racing car in 1951. The designers called their prototype the Disco Volante (Flying Saucer) for it was practically symmetrical about a horizontal plane—a full body rather than a half-body. The name was more appropriate than they guessed, for the car had horrible lift problems. It was a costly lesson, and it took Alfa Romeo almost 10 years to master the problems of streamlining a competition car. When it finally happened, it was done with the collaboration of Carrozzeria Zagato in Milano: the Giulietta GTZ of 1962.

Alfa Romeo chassis had been used for two most interesting streamliners built by Carrozzeria Bertone, to designs by Franco Scaglione. He was an industrial designer who had

Bertone's BAT-5 of 1953 introduced a series of experimental streamliners using Alfa Romeo chassis. All wheels were enclosed to below hub level and tailfins curved inwards.

learned more than the rudiments of streamlining, for his BAT-5 of 1952 and BAT-7 of 1954, were bodies of extremely low drag. He had first created attention with a spectacular Abarth 1500 prototype in 1950, and the BAT models (BAT = Berlina Aerodinamica Turismo, which means something like streamlined closed touring car) were indeed more scientifically designed than any of the Chrysler-Ghia projects. Scaglione used curved tailfins to guard against lateral separation along the tapered roofline, and paid extremely close attention to air intakes and outlets.

Scaglione went on to design a series of speed-record cars for Abarth, most of which were built by Pininfarina, some using Alfa Romeo chassis, and others being based on Fiat underpinnings.

The leading aerodynamic scientist in Italy was Dr. Alberto Morelli of the Torino Polytechnic Institute. He concentrated his studies on the small sedan, and came up with several remarkable proposals, which were realized by the cooperation of Fiat and Pininfarina. After Fiat had launched the rear-engined 600 Multipla (which was a forward-control van-type car, popular in taxi service), Morelli designed a highly streamlined version.

In 1960 he went one big step further, placing the four wheels in lozenge formation, which enabled him to reshape the

The BAT-7 was a logical evolution from the BAT-5, going further in the direction of an aircraft-type nose, and raising the tail fins for greater effect. Hot under-hood air escaped at the cowl sides and the scoops at the head of the rear fenders admitted cooling air for the brakes.

body with a more pointed nose and a longer, more smoothly tapered tail. It was Morelli who laid out Pininfarina's full-scale wind tunnel, completed in 1975.

British Development Came Through Racing

On the competition front, British constructors had several contributions to make. First, the Bristol 450. In 1953, the Bristol Aeroplane Company built a team of cars for the 24-hour race at Le Mans, and as befits an aviation firm, the bodies were smoothly streamlined. However, they appeared at Le Mans full of holes and louvers, for during pre-race trials, Bristol had suffered severe cooling problems. A basic shape with low drag is no good unless the internal airflow is included.

Appearing in 1955, the BAT-9 was almost a prototype for the Alfa Romeo 2000 that was produced in very small numbers in 1956. Traditional Alfa grille was restored and tailfins were de-emphasized.

Pininfarina and Abarth collaborated on a series of speed-record cars in the 1950s. The shapes were developed for minimum drag with little regard for lift and stability problems. They had small engines and did not reach speeds where those factors were critical.

The car had a narrow cockpit on a wide pontoon, with a long hood, and a roofline with full-length fins.

The body panels were considerably cleaned up the following year, and the Bristols went faster. However interesting this experiment may have been, it really led nowhere, for the production models followed contemporary fashion, and nobody else copied the Bristol 450 style.

The name Jaguar is automatically brought to mind whenever Le Mans is mentioned, and the aerodynamic aspects of Jaguar's performance on the Sarthe circuit are most illuminating. They started with the C-Type of 1951, which was about as wind-cheating as the Bugatti and Alfa Romeo racers had been in 1937. But Stirling Moss set the fastest lap with the C-type, so Jaguar knew it had race-winning material. At the time, Jaguar had a staff aerodynamicist named Malcolm Sayer. He did not have access to a wind tunnel, nor did he much believe in wind tunnel testing. He worked it all out mathematically. For 1952, he calculated some improvements into the C-type body, giving it a longer and flatter nose, which he thought would raise top speed enough to make it the fastest car on the track. A team of three cars were sent, and all were eliminated within a few hours, due to overheating.

In his calculations, Sayer had overlooked something important about interior airflow. He did not recover from this setback for a long time, and for 1953 Jaguar raced the C-type with the 1951 body. As his revenge, Sayer produced the D-type for 1954. Here was a potential flying saucer, with the lift problems effectively doctored out of it, and a big tailfin integrated with the fairing for the driver's head rest to keep the car on course. It had not been wind-tunnel tested, but derived from mathematical studies. A remarkable result, in view of the mistakes still being made by other companies that did use wind tunnels but erred in translating the test data.

Two other British sports cars were to have a far-ranging impact on competition-car streamlining. First came the Cooper 1100 in 1955, with a midships engine, and a new version of the Kamm-heck, with a concave panel at the tail end of the cut-off body. Next came the Lotus Eleven in 1956, with a very low and flat body, on top of which the fenders, headrest, and windshield were faired in.

Citroen Leads Modern Mass-Producers

At this point we must return to production cars. There was a milestone car that made its debut at the Paris auto show in

Appearing in October, 1955, the Citroen DS-19 was a mind-staggering demonstration of how to apply streamlining to a five-seater sedan. The body remained in production for 20 years.

1955 and was to remain in production for over 20 years: the Citroen DS. The design was credited to a sculptor, Bertoni, who said his inspriration for it was an aquatic fowl. But of course, the actual dimensions and proportions were determined by Citroen engineers, who also had to decide on structure, materials, manufacturing and assembly techniques.

The DS was a perfectly rational design, and consequently one that showed no concessions to fashion. Many described it as weird, but it was both functional and aerodynamically efficient. Since that time, however, Citroen has been working with a design consultant, Robert Opron, who was largely responsible for shaping the SM (1970), GS (1971) and CX (1974).

The Triumph of Negatives

While touring-car design evolved along the paths I have described, new aerodynamic devices were beginning to appear in the racing car field. In 1961, Ferrari used a "ducktail" spoiler on the deck of the midships-engined 246 open roadster, and in 1962 a similar ducktail adorned the lower lip of the fastback roofline on the 250 Berlinetta GTO.

The front spoiler was mainly a Ford development, proving its value on the GT-40 in 1964. But rear spoilers were to have quite a history. Their history can actually be said to have started with the wings on the 1928 Opel Rak-2, since their task was to reduce lift.

Negative-lift airfoils, short and shapeless, were also a feature of the vehicle Mercedes-Benz built for an attack on the world land speed record in 1938. The car never ran, but the

Stuttgart company went to the trouble of patenting the wing design. And Mercedes-Benz continued to study the use of aerodynamic devices.

A roof-mounted wing installed on a Mercedes-Benz 300 SL during practice for Le Mans in 1952 was nominally in air brake—intended to slow the car by tilting it vertically prior to entering slow curves. Remember, the 300 SL had drum brakes, and in a 24-hour race it was important to avoid relining. The device was not used in the race, however. But three years later, a similar air brake, hydraulically actuated, was a feature of the three 300 SLR machines entered at Le Mans. The cars were open, the "wing" normally blended in with the sheet metal on the rear deck, but could be raised to a near-vertical position. Fangio and others discovered that leaving it up through the curves rather than folding it as soon as they had lost enough speed, had the effect of improving traction and stability. As you can see, it worked as a negative-lift device.

From Chaparral to Stock Cars and Corvettes

You know about the much-copied Ferrari ducktail. Beyond that enters the name of Chaparral, a racing car built in Midland, Texas. Its designer, Jim Hall, fitted a driver-controlled airfoil between the finned rear fenders of his 2C in 1965. It worked very well, reducing lift and adding traction whenever needed, and encouraged Hall to experiment further with wing structures. One year later he installed high-mounted wings, carried on vertical struts, clear of the air flow around the body, on the Chaparral 2E.

When wings were regulated out of usage, Hall developed the negative ground-effect machine, or the vacuum-cleaner, as it was less respectfully called. Its official designation was the 2J. How did it work? Just the opposite of an air-cushion vehicle. Consequently, it looked a lot like one, with aprons hanging down from the body sides to the ground, to minimize air leaks. But instead of pumping air from outside to build up an air cushion under the car, the Chaparral pumped the air out the back from under the car, via two big fans driven by a separate engine.

The idea was to generate a negative lift force in the form of suction. The fans were capable of producing an 1100-pounds' depression under the car, which gave it astounding adhesion on curves. It was effective at all speeds, whereas wings work

Chaparral "vacuum-cleaner" car. Flexible skirts sealed off the car's underside, and motor-driven fans pumped the air out to create suction conditions (vacuum) under the car.

best only at very high speeds. The fans, aprons, motor, and other accessories did not add to the total drag. The 2J was built in 1969, and remained on test through the 1970 Can-Am season. Then the FIA declared it illegal. It became an instant museum-piece and Jim Hall never built another Chaparral.

Stock and Sports Car Racers

By 1969, winning stock car races had become a matter of the utmost importance to Chrysler, whose engineers realized full well that more than half of the engine power was being used to overcome aerodynamic drag. A program to develop a streamlined stock car racer was started and assigned to R. P. Marcell and G. F. Romberg, who went to work with a 3/8-scale model in the wind tunnel at Wichita State University (which has a 7 × 10 ft. test section) and at the Lockheed-Georgia Company in Marietta (which has a 16 × 23 ft. test section). Starting with a standard Dodge Charger, they created the

Daytona model by fitting a curved and elongated nose section and adding a front spoiler, raising the rear window angle from 45° to 22° from horizontal, adding a 1-inch deck spoiler plus a 58-inch wide wing carried on struts 23.5 inches above the deck, and adding tailfins on the rear fenders.

The result was a 15-percent drag reduction, and a complete change in the vehicle's lift pattern. Front end lift at 200 mph was brought from a high positive value into the negative area. Finally, the Daytona had far greater resistance to sidewinds. A limited series of Daytonas was produced, but the lessons fom the program have not been incorporated in Chrysler's mass-produced cars.

Just as it's easier to style a sports car than a family car, it's easier to streamline a two-seater than an economy sedan. With the possibility of low seating and unrestricted length, the designers can unbridle their imagination. But when carrying capacity, size and weight have priority over streamlining, the result must be a compromise. Citroen's GS remains unsurpassed as the example where streamlining has been the least compromised.

How the sports car can be streamlined depends to a large extent on where the engineers put the engine. With a front-mounted engine, the car is bound to have (1) a hood of some length, (2) front air intakes for engine cooling, and (3) a higher roofline as a result of the drive line to the rear wheels and the impossibility of lowering the seats as much as in a midships- or rear-engined car (except in cases when front wheel drive is adopted).

The Chevrolet Corvette has remarkably good streamlines for a production car. The shape was evolved in wind tunnel testing, and is worthy of attention. What the Chevrolet styling studio is capable of when the engine is placed behind the seats

Dodge Charger Daytona obtained higher speeds through drag reduction—streamlined nose. Lift forces were overcome by rear wing structure.

Ford went after the Corvette market with the midships-engined De Tomaso Pantera. Its failure in the market is no reflection on its aerodynamics, which are excellent.

is shown by two experimental Corvettes exhibited in 1973. Why they were to remain experimental is difficult to understand, especially as Ford was at that time producing the Pantera in Italy, with a very low-drag Ghia body on a midships-engined chassis.

Some Extreme Accomplishments

Pininfarina has demonstrated the ultimate in streamlining a four-seater with front engine, in the Studio Cr-25 of 1974. It has a drag coefficient of 0.256, and its control of interior airflow is particularly noteworthy. Cooling air for the engine is scooped up below the bumper and exits through three slots on the front part of the hood. The bumper is shaped as an airfoil, to reduce lift forces acting on the front. The horizontal louvers in the rocker panels and door sills provide entry for fresh air to cool the exhaust pipes. Curving tail-fins promote high-speed stability, and a fixed wing between them does double duty in maintaining laminar flow across the rear window as well as acting as a negative-lift airfoil.

With centrally mounted engines, the hood can be eliminated. That is, it can take the form of a panel that extends the windshield plane to the very nose of the car. Two such examples stand out, the Ferrari 512 S of 1969 and the Lamborghini Bravo of 1974. The former was a Pininfarina creation; the latter was a Bertone product. Both are noted for their clean lines—a total lack of fins, spoilers, wings and air dams. Both have a wedge theme—unbroken on the Bravo, and sort of in two stages Ferrari.

Both had front radiators, with thin slots in the nose serving as air intakes. On the Bravo, breathing air for the engine entered through ducts in the rear quarter-panels, while the Ferrari had big ducts in the door panels on both sides of the windshield, with an exit path through the louvered roof.

Pininfarina's Modulo of 1970 was also built on a Ferrari 512 S chassis. It revived the flying-saucer theme, and was in fact nearly symmetrical about its belt line. It could also be described as a flattened-out version of the Goettinger form of 1938. Cooling air was admitted through a slot-type air intake, with NACA ducts on both sides for brake-cooling and ventilation air. All wheels were enclosed, but an opening was provided for the protruding top of the front tires, and a faired-in fender shape emerged from the basic shape to cover the wide rear tires.

It is not difficult to streamline a two-seater coupe. It can be done well in several ways, as we have seen. But to streamline a four-door sedan, especially an economy sedan, now that is difficult. Streamlining and making it look good at the same time is almost impossible.

And that's where the future is. The economy sedan seems destined to be the average car from now on, and that means all the mass-producers will be obliged to come forth with their own particular solutions.

The basic knowledge is there. What hinders and what helps the air flow is well understood. The main problem is to get the good ideas out of the wind tunnel and on to the assembly line. In the past, the auto industry could afford to ignore streamlining. That will no longer be possible.

chapter 7
car architecture and streamlining

Vehicle architecture may sound like a hifalutin' way of saying design, and in a way that's true, but I am using the term in order to distinguish design at the engineering level from design at the styling or cosmetic level.

Space utilization is a primary consideration in vehicle architecture. That means providing a maximum space for passenger accommodation within the selected dimensions—and ideal dimensions are always restricted. With smaller size, the weight of the car is kept down, which reduces the amount of raw materials that go into it, and the amount of energy consumed in manufacturing it. Lower weight is also an essential element in obtaining improved fuel economy.

Smaller overall size will usually also lead to a reduction in frontal area—with benefits in terms of diminished air drag.

The designers—or vehicle architects—have countless decisions to make. Where to put the engine? Seats for two or three abreast? Drive the front wheels or the rear ones? At the end of the list comes the clincher: Which decision to make first?

Carrying Capacity Comes First

Naturally, you cannot start off with the C_D and force all other parameters to fit. The only logical approach is to start with the payload. The car is a transportation machine, and all other design objectives must be subordinate to its carrying capacity.

Adapting the SM lines to the body of a short family car was a tremendously difficult task, but Robert Opron and the Citroen design staff succeeded with the 1971 GS.

Volkswagen Microbus is an outstanding example of how to maximize useful space within compact overall dimensions. But it has large frontal area and a high aerodynamic drag coefficient.

High carrying capacity does not necessarily enter into open conflict with streamlining. It's possible to design six-seater sedans with outstanding streamlines. But if the design goals also set limits on length and weight, it is obvious that some compromise must be made. And then the designers must decide whether to cut back on seating capacity, or to rob the passengers of some space and comfort, or to accept higher air drag.

If your objective is to provide seating for nine on a 94.5-inch wheelbase, you are looking for an extreme solution, where conventional thinking must be forgotten. Perhaps you think it isn't even possible. Well, Volkswagen did it, and kept the C_D down to 0.40. How? You've seen the Microbus, of course. It's shaped like a shoebox, without a hood, and has a rear engine. It has other aerodynamic drawbacks, such as large frontal area, and low directional stability in side-winds. But there are numerous lessons to be learned from this design, which dates back to 1949.

Lesson Number One is that a box is roomier than an eggshell. The problem with that is that the eggshell has much lower drag as it moves through the air. And here we are back to the basic payload question again. Less room, lower carrying capacity.

If you will reduce your payload requirement, you can begin to move away from the box towards the eggshell shape.

Remove the back seat, and you are suddenly designing a completely different type of car. Streamlining becomes easier as your payload is reduced.

But that does not mean you cannot make the most of streamlining the shoebox when the specifications call for a lightweight family car with 4 to 5 seats, and a trunk of at least 12 cubic feet useful space.

Alfasud: Logical Analysis of Details

An excellent example is provided by the Alfa Romeo Alfasud four-door sedan. It creator, Rudolf Hruska, explained to me the thought processes that led him to the design formula that was adopted. As he is about six-foot-two and test-drove the prototypes for more miles than anyone else, the seating problem would not be neglected, but in fact was taken extremely seriously, and given first priority.

The driver's seat must have 6 1/2 inches of fore-and-aft adjustment, there must be at least 23 inches between the pedals and the lower end of the steering wheel, and 16–22 inches between the backrest and the steering wheel, according to seat position. Headroom was set at 38.5 inches in front and 36 inches in the back. The open space between the back of the front seat and the rear backrest, at hip level, must be no less than 27.5 inches, with the driver's seat in its rear-most position. For minimum acceptable seat height, he chose 12 inches in front and 13.5 inches in the rear for the vertical distance form the front edge of the seat cushion to the floor. Door-to-door interior width was set at 54 inches. Except for less width, these dimensions compare with some American intermediates of the 1968-1975 period.

As Hruska imposed the rule of containing the rear seat entirely within the wheelbase and would not accept any noticeable wheel housing intrusion into the pedal area, this necessitated a wheelbase of 96.65 inches.

The next decision was where to put the engine. Hruska had worked with Porsche for many years, and was familiar with the rear engine layout. During his seven years at Fiat, he was mostly associated with front engine/rear drive models. But for the Alfasud he chose front wheel drive.

The reasoning was purely logical. The engine must be at the end where the driving wheels are, because that gives the best traction. It eliminates the propeller shaft which is a

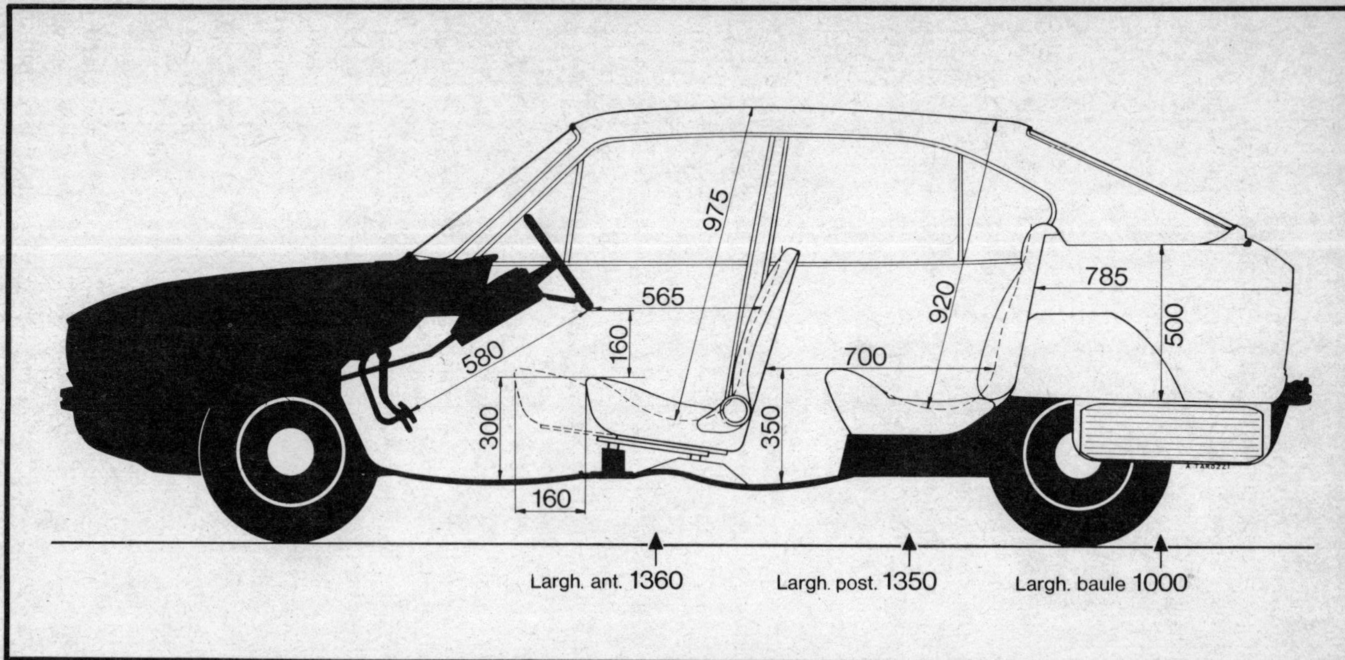

Space utilization is brought beyond all accepted norms in the Alfasud sedan. Mechanical elements take up very little space, and are concentrated in the front end. Streamlining is very good, too.

possible source of vibration. Integrating the entire power train also reduces production cost.

The question of whether to put everything in the back or in the front was resolved by considering the small trunk spaces available with rear engine design, plus the near-impossibility of making a station wagon version with that layout.

Front wheel drive, on the other hand, offers better trunk space in the tail, and a low, flat floor for a wagon. Front wheel drive also automatically provides high directional stability, and assures freedom from handling and roadholding problems.

So far, so good. Now for the engine. What type and what size? Hruska wanted a short hood and a low hood line, and the engine configuration which combines the lowest height with the shortest length is the flat-four (horizontally opposed, with two cylinders on each side). It is inherently better balanced than the in-line four-cylinder engine, assuring less noise and vibration. This type of engine also has a very low center of gravity, which contributes to better ride and handling. Water cooling was chosen, for greater silence and a more efficient heater, which necessitated finding space for a radiator in the nose section.

As all space aft of the front wheel axis had been taken up, the engine was placed outside the wheelbase, with the final drive unit in line with the wheel hubs, and only the gearbox extending backwards into the floorboards.

Engine displacement was determined by the performance objectives. When the design work began in 1968, Hruska aimed at 55 hp (DIN) but later realized that higher performance would be good sales argument, and upped his target to 63 hp (DIN) or 73 hp (SAE). These targets were reached without raising the initial 73 cubic-inch displacement. Specific power output in this range is normal for Alfa Romeo and is reached without increasing the internal stress levels to the point of threatening reliability or durability.

The little engine weighs only 240 pounds complete, so that it does not cause any weight distribution problem—as might be the case with a heavier power unit in the same position.

While Hruska and his assistants were engineering the mechanical elements of the Alfasud, Giorgetto Giugiaro of Studio Ital Design was engaged to design the body. He was given an overall limit of 13 feet in length, and the minimum

No boxiness is apparent in the Alfasud body which has a well-rounded nose and a fastback roofline cut off sharply at the belt line.

headroom dimensions which would fix the roofline, plus a generous trunk space requirement. He was instructed to provide maximum visibility in all directions, which meant a low belt line and big glass areas. Also the doors should make full use of the space between the wheels for easy entry and exit. Giugiaro pushed the windshield far forward, and made it a flat center section, with well-curved corners. He chose a rear end design that's a fastback down to the belt line and then becomes a vertical trunk lid. Fuel tank and spare wheel were to go below the trunk floor. Giugiaro managed to get a trunk floor length of 31 inches and a maximum cargo height of just under 20 inches. Out of the car's 13-feet length, 9 1/3 feet were devoted to the passengers and their luggage. The final design had a C_D of less than 0.30, which is remarkable for a car of this class.

With a curb weight of only 1835 pounds, the Alfasud had a top speed of 95 mph and gave 30-mph gasoline mileage in everyday driving. The body shape avoids the boxy look completely, though the space requirements clearly indicated that only a box could do the job. In contrast, designers who stuck to the box concept (or double-box, if you prefer, regarding the upper structure as a small box on top of a big box) have achieved miracles of space utilization but at a cost of disregarding air drag. Typical examples are Fiat 128 sedan, Fiat 124 sedan, and the Volvo 144 sedan. Mercedes-Benz and BMW are not quite so good on space utilization, and somewhat better on aerodynamics.

Poor Space and Weight Control in America

In America, the best examples of space utilization have been the Plymouth Valiant and Dodge Dart from 1966 to 1976, but they had body shapes of relatively high drag. During that ten-year period there was practically no control over weight for large American cars, in spite of which the designers still came up with small, cramped and uncomfortable interiors (witness the Thunderbird, Continental Mark IV, and Pontiac Grand Prix).

For 1977, GM put all its big cars on a quick weight-loss diet, and cut overall dimensions by about ten percent, while maintaining or enlarging the interior space. This program will continue through all car lines, and other manufacturers will have to follow.

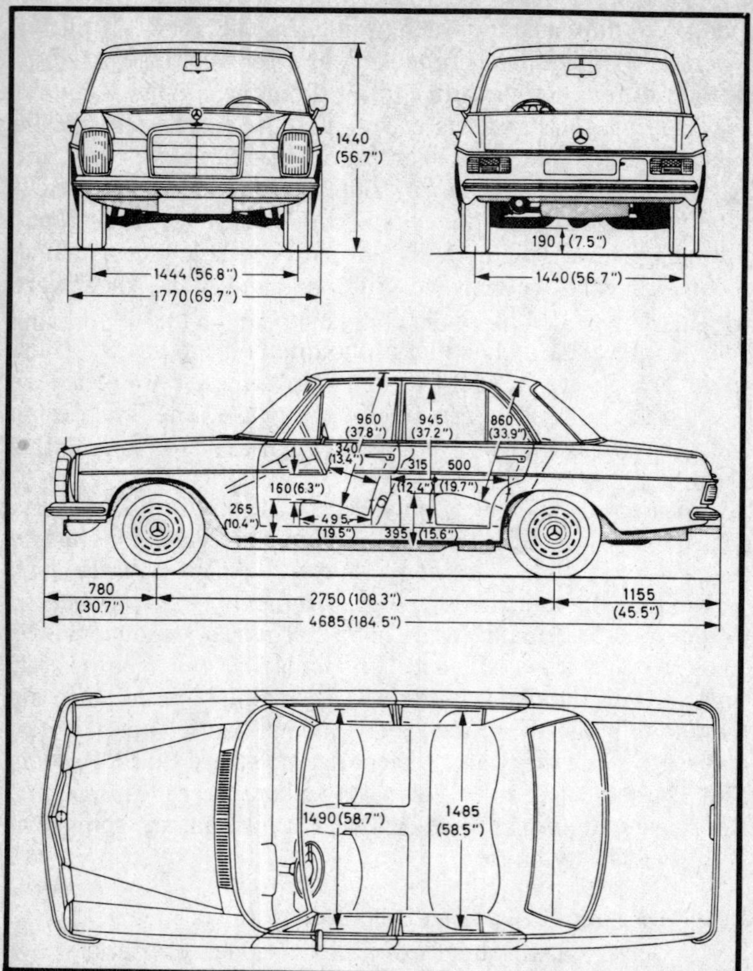

Mercedes-Benz body is typical of box-on-box layouts, which offer good space utilization, but gives a shape that is difficult to streamline.

Weight control is an inseparable part of vehicle architecture. It provides a ceaseless incentive to reduce mass and size. Yet many auto manufacturers do not practice weight control on a top-priority basis.

Manufacturers differ in their approach to use of sound-deadening materials, sealants, weather-proofing, insulation, upholstery and interior trim. Cars of similar size can vary considerably in weight. Sometimes high weight means heavy frames and thick sheet metal. But it can also

The ultimate in family-car streamlining? It is difficult to see how the Citroen CX body design could be improved. Thanks to low drag, the four-cylinder engine gives it a top speed of 120 mph.

mean that weight control at the design stage has been poor. Unless there is a conscious weight-control program, with strict allocation of responsibility to meet targets in each area, you will find parts carrying excess metal which contribute nothing to strength, structures with heavy reinforcements that could have been avoided by proper stress analysis at the initial design stage, and weight that's added to combat some non-structural problem such as noise and vibration.

Weight added by streamlining efforts comes in a special category. As long as we are talking about everyday sports cars and family cars, I feel there is no such thing as a legitimate growth in weight due to aerodynamic design. Devices added to the finished car may add weight (spoilers, air dams, etc.) but that's because they were not part of the basic design. If they had been, some weight could have been eliminated elsewhere. For racing and speed-record cars, the situation is quite different, both with regard to the need for negative-lift devices, stabilizing fins, and with regard to weight, for the racing formulas stipulate minimum or maximum weight limits for various categories and groups.

chapter 8

stylists vs engineers

Ideally, the styling center of an American car company is where art meets engineering and blends into the creation of new shapes that bring visual enjoyment to the viewer. In reality, styling has been a tool for the industry to make old cars look new and attractive.

For half a century, Detroit's auto stylists have done more to repulse the advance of streamlining than to make use of the aerodynamic knowledge available to them. Why this rejection of natural laws? Why this resistance to make use of a science that could—if properly applied—give some new model a true competitive advantage? The answer, in one word, is fashion. Not fashion in itself, but the car designers conception of what a fashionable car should look like. When styling began, in the late 1920s, fashion called for long hoods and tall radiator grilles. The idea was to create an image of power and speed, but curiously the stylists chose their inspiration in the steam locomotive of the past rather than in the airplane of the future.

Chrysler's experience with the Airflow told them they were right. Styling was an instrument for the sales department, not a workshop for making the car slip through the air more efficiently. This situation was not just something that evolved to fill a void. It was a consciously pursued goal on the part of the industry. Some quotes from Harley J. Earl, who was the founder and head of the GM Styling Staff for about 30 years, will illustrate this: "By creating new features and new

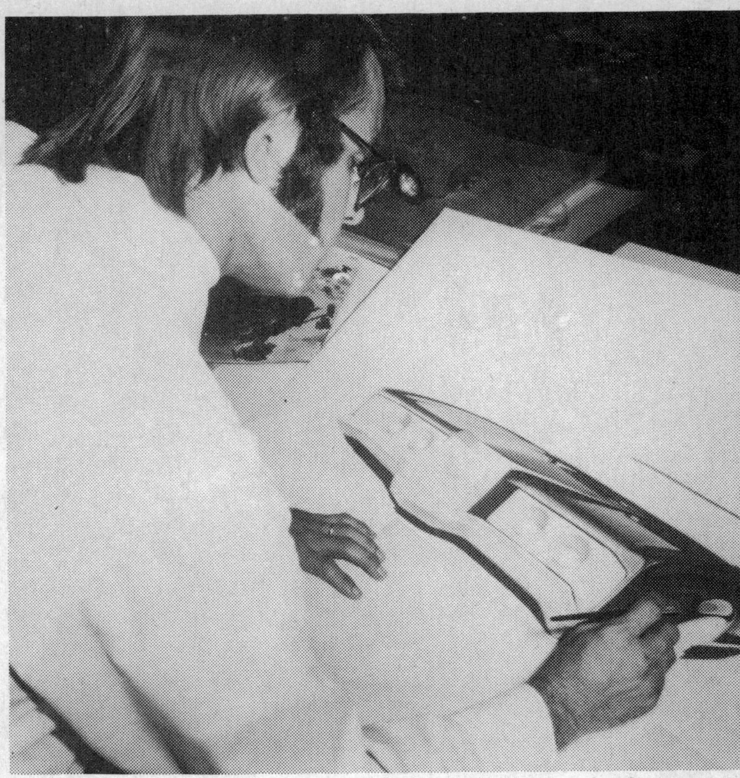

Frontal appearance has always been the uppermost concern of American stylists. This sketch for a proposed Mercury shows a total disregard for streamlining.

products, it is the hope of General Motors each year to render obsolete those products which have come before. Thus, the design of the car can be a persuasive tool for inducing car buyers to get rid of the old and buy the new."

In his search for innovation, Earl was guided mainly by market research. In other words, by public opinion, as opposed to expert knowledge. GM used to exhibit far-out dream cars, shapes that took no account whatever of the actual traffic picture or paid much attention to the car's basic function as a conveyor of a payload. "Public reaction to these cars is carefully tabulated and the results heeded," said Earl.

Engineers Oust Stylists

This has now changed to a large extent. The science-fiction flavor, the hot-rod gimmicks, the psychedelic clamor for

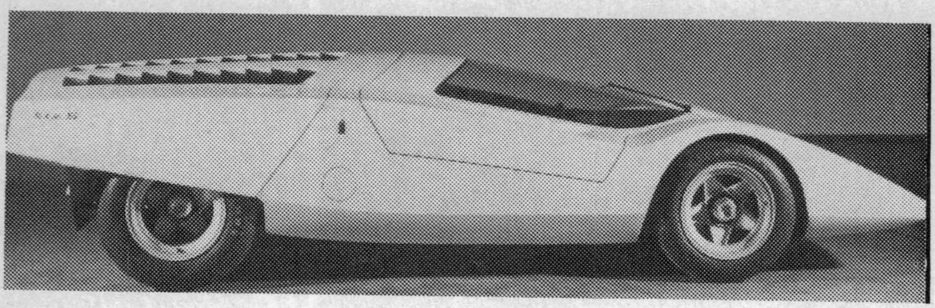

Angular difference between hood line and windshield has been eliminated from Pininfarina's Berlinetta Speciale on Ferrari 512-S chassis. Engine cover continues roofline at high level to the very end.

attention, all have been played down. More and more stylists also have some engineering training. At Ford, the styling center was renamed "design center" some years ago, because styling had become a bad word—associated with a lacking sense of reality, lacking concern for the consumers, but endowed to excess with bad taste.

Engineers turned out to have a better idea of future conditions and future demands than the stylists. Back in January, 1959, Jack Charipar, who was then chief engineer of Plymouth, told his audience at an SAE meeting in Detroit: "By 1980, automotive styling and buyer taste will have outgrown present preoccupation with size and mass as ends in themselves to concentrate on better use of space and materials." Mr. Charipar also cited aerodynamics as coming into many more practical applications, to make "the car of 1980 more effective in its use of fuel, quieter, and more agile at any speed—and safer."

It's all coming true. The auto designers are now working closely with aerodynamicists, and their ideas are wind-tunnel tested. The traditional downplaying of true streamlining in America is at an end.

There has never been any factual conflict between aerodynamics and styling, as evidenced by the work of countless Italian coachbuilders. Most of them began by building bodies for racing cars, and they made them as streamlined as they knew how. Whether they won or not, they set the fashion. This trend helped the advance of true streamlining in Europe, from racing prototypes to limited-production sports cars, from Gran Turismo to low-priced sports cars, and from prestige limousines to family sedans.

Now the auto makers all over the world are trying to streamline better, and the question is what this will do to styling. Will it be the end of styling? Will all cars come to look alike?

The answer is, of course, that cars will continue to be different, but in new ways. There are many reasons why cars will always come in a variety of types and styles. They are not all designed for the same purpose. They vary in seating layout,

Full-scale models for Pontiac show various grille and headlight proposals for Bonneville and Grand Prix. They show strong Pontiac identification, but oppose the basic rules of streamlining.

engine position, chassis configuration, and dimensions. Within some categories, there may be a tendency for cars to adopt the same general shape. This means that the stylists will gain in importance, for it will be their task to give their model distinction, and to make it better looking than rival models.

Introducing Designers to Aerodynamicists

Streamlining can also provide a tremendous stimulus to designers. While it restricts them to efficient aerodynamic shapes, they will have greater freedom to, or indeed come under increased pressure to, create new types of body. We have seen the hatchback vogue flourish, and more and more manufacturers are turning to the idea of a sportswagon (i.e., a wagon based on a sports car rather than on a family sedan).

With the demise of the convertible, there is room for lots of new ideas to give people back some of that "outdoors" feel. Designers may find their answers in recreational vehicles, for instance, or invent totally new combinations that will increase the consumers' choice.

But the stylists will have to learn to work with aerodynamicists just as they had to learn to work with engineers. This can happen smoothly, as long as there is a project director who understands that streamlining must have priority. Stylists must be ready to revise their shapes if wind tunnel tests show they have problems. There are many cases where that has been done. For instance, Volkswagen commissioned Giorgetto Giugiaro of Studio Ital Design to "clothe" the Golf (Rabbit) and Scirocco. He delivered the designs and prototypes, but VW put them in the wind tunnel before clearing them for tooling. The result was that a number of small problem areas were found, improvements devised, and changes made. That's the stuff that better gas mileage is made of.

Take the case of Mercedes-Benz. Here, styling is just one of the many departments in the body engineering section. There is no director of styling, but a small group that works under orders from the technical department. The engineers supply the basic package, with all dimensions settled within close tolerances. That doesn't mean there is no styling left to be done. On the contrary, the task of styling a compact sedan is extremely challenging. You must work with great subtlety to illustrate the car's character on nothing more than a thin

Full-scale clay model for Ford Torino is shown in open air at Ford's design center in Dearborn. Profile shows streamlined basic shape, but accent lines will create airflow disturbances.

skin around a box (or a box-on-box, in the Mercedes-Benz example). You cannot add five inches to the nose just because you think it makes the car look better. Because the easy solutions are outlawed, the stylists have to extrapolate their imagination, and bring every ounce of their talent to play on creating a visual impression that makes a statement about the car. This is difficult. Not only that, but at Mercedes-Benz, tradition must also be observed. The new model must not break with past styling themes, but appear as a continuation of the same theme.

Some companies do not have this concern. Volvo, Renault, Lancia, Rover, MG and Jaguar are examples of this. Their designers jump all over the map when it comes to lines, curves, proportions, and overall images.

Company Continuity and Styling Cycles

At BMW, Alfa Romeo, Porsche, and Opel there is a studied continuity between each new model and its predecessor. That's too strong a statement to apply to Fiat and Peugeot, though both have a definite family resemblance throughout their model ranges, something indefinable that helps identify the car as a Fiat or Peugoet by its form, long before you are close enough to read the labels.

In America, General Motors has always cultiviated a certain continuity. This was broken up to a large extent during the model proliferation of the 1960s, as cars that had no heritage burst on the market every so often: Corvair, Chevy II, Rivera, Cutlass, Tempest, Toronado, Monte Carlo, Camaro.

Chrysler occupies the middle ground between General Motors and Ford. It's Ford that has the wildest swings in styling, between one car that's going out of production and its immediate replacement. It's obvious that Ford has no general styling policy, and there isn't even a family resemblance through the model range.

The U. S. industry works with a 3-5-year styling cycle, which means that a new body will run for three to five years with annual facelifts. Then it's replaced by a direct successor. In Europe it's much different. New models are created for production lives of 8 to 20 years. Some don't make it over the full scheduled period due to lack of demand, while others go beyond. When they die, many are not replaced, but instead the lineup is juggled so that the gap is filled by other models, from below and from above. This is actually more a matter of merchandising than of styling, but it shows how family resemblance between different models can benefit a manufacturer.

The stylists do not as a rule dream up new projects as they please but work strictly to order. The order comes from a committee of top executives, market researchers, and product planners. The engineers get a similar order. Sketching begins. Scale models are made. Some are just looked at—and some go for wind tunnel tests. Another group will tackle the seating, while body engineers prepare a preliminary structure. The sketches are replaced by full-scale exterior layouts. A seating

Early proposals for the Chrysler Cordoba show considerable respect for aerodynamics, with flush windows and no front corner posts. But final production car has mediocre streamlining.

buck and a full-scale clay mockup are produced. Interior layout, instrument panel design, and interior trim are developed. Body engineers work on fabrication and interchangeability studies. A weight analysis is made. Full-scale wind tunnel tests are performed and evaluated.

Throughout this phase, the project remains flexible in styling terms. The market may send new signals, that call for a change in the product. But at some point, the committee will approve the car for production, or kill the project. If it gets the green light, prototype construction begins. The styling job is finished.

chapter 9
streamlining and safety

It is not because safety is the lowest in priority that I have waited to deal with it last. Safety ranks high in importance, but it's extremely difficult to discuss until all other design objectives have been examined. Safety is very difficult to discuss at any time, for it's one of those subjects nobody is against. Everybody is for it—but there are limits to what it must cost. I am not referring strictly to purchase or running costs, but also to what might be traded off against more safety.

Safety features must not alter the basic qualities of the machine. Efficiency, performance, versatility, maneuverability, comfort and convenience are fundamental items in buyer motivation, and each is threatened by some aspect of making cars safer.

Streamlining is not necessarily in jeopardy because we are concerned about safety. Very often design elements produced in the interest of safety lead to better streamlining, and vice versa.

But in all discussions of safety, it is crucial to define what we mean. There are not many points where streamlining and safety meet, so it is all the more important to identify them and examine the conflict or interaction.

Regulated Safety and Real-life Safety

First, we are able to zero in on the vehicle itself. The two other broad areas of safety—the driver and the environment

Highly streamlined SNU Ro-80 body shell undergoes crash test, mounted on a sled. Streamlining and safety can go hand in hand, if the principles of both are fully understood.

(roads, signals, traffic) can be ignored with regard to aerodynamics. In the vehicle itself, we have two modes of safety: Active and passive. Active safety relates to accident prevention: driver action and vehicle response. Passive safety relates to occupant protection when an accident occurs.

Our concept of safety also holds a double meaning. There is safety as defined in the federal motor vehicle safety standards; and there is safety under real-life conditions. The basic problem with safety standards is that there is no such thing as a standard accident. If there were, and people would only have standard accidents, you could eliminate traffic fatalities. Consequently, whether a safety standard does any good or not, depends on how closely it conforms to real-life conditions.

We do not have to look far to see how standards can get into a fight among themselves. Clare MacKichan of GM Styling said in 1970: "The position of the lighting elements is now rather strictly prescribed. The requirements for cooling the engine add to their demands, and increased bumper protection may find itself in conflict with these requirements. This now says to the designer: 'You must exercise your creativity within narrow limits. Your park/signal lights must be outboard; your headlights must be 24 inches off the ground and must be visible from a prescribed cut-off angle. Your bumper protection must be extended over a vertical range, and you'd better not forget to leave enough holes in the front of the car for cooling the engine'...."

When you are talking about bumper protection, you are not discussing auto safety in loss-of-life situations. You are talking about protection, not for the occupants, but for some of the vehicle's safety systems, such as lights. But we have bumper standards, and while they may be said to be an obstacle for progress on streamlining, they could also work as an incentive to create new shapes and systems that would bring aerodynamic improvements in the long run.

On the other hand, it's vital to the cause of safety, in real-life terms, that streamlining must assure adequate directional stability at speed, and that streamlines must not present a danger to others—car occupants, pedestrians, cyclists, and others. In these areas, the industry is still working in complete freedom from standards. In the first instance, streamlining can play an effective part in active

"Safety-car" concepts such as this Fiat ESV-2000 contain ideas that are often detrimental to air flow. This front end is sadly un-aerodynamic with the recessed grille conflicting with the bumper system.

safety. In the second, you can look at diminished agressiveness as a foreign element in passive safety.

From the Driver's Seat

Streamlining has another contribution to make in active safety: improved driver visibility. A low belt line, sloping hood, and large glass area are essential to this end, and tend to lead also to better streamlining. Outside rear view mirrors, on the contrary, are a legal requirement (and indispensable to traffic safety) that only cause aerodynamic problems.

We may also have a small problem with the glass, especially the windshield glass. Laminated safety glass is a legal requirement, and it can be shaped with compound curves to fit the aerodynamic pattern perfectly. But curved glass also means there will be some distortion, and that can be a threat to safety.

On the standards front, the National Highway Traffic Safety Administration is trying to formulate rules for maximum allowable obstructions in the driver's field of direct view; light-transmittance levels of glazing materials that provide fields of direct view; visibility of the vehicle corners from the driver's seat; and view interception, adjustment, and light transmittance of sun visors. US federal motor vehicle safety standards permit concealed (pop-up) headlights, on condition they are fail-safe in terms of never closing or shutting off the light inadvertently. Concealed lamps can be hidden under flush-fitting covers on the hood or fenders, and minimize an important source of drag. But the American law does not yet allow headlights that are recessed in working position under a plexiglas or other transparent cover that blends smoothly into the body skin. European companies like Cibie, Marchal, Carello and Lucas have developed headlamp lenses in a variety of shapes and sizes, and work closely with the manufacturers to provide units that fit the shape of the body with the least aerodynamic disturbance.

Streamlining and Crash Survival

Finally, we come to the deadlock. It's the ultimate question of passive safety: crash survival. Streamlining may have a big part to play here, as the envelope for the structures needed to absorb the energy of the impact may be quite different from the envelope that simply encases the

Hydraulic bumper system on AMF experimental safety car looks as if it's meant for a railroad locomotive and not for an automobile. It could be streamlined, by adding a C-shaped front of rubber or plastic materials.

mechanical elements of the car. Both dimensions and shape will be affected, and in the case of frontal and rear impacts, the structural reinforcements can indeed prove to work in favor of streamlining.

Statistically, the frontal impact is the worst killer. But the offset vehicle-to-vehicle frontal impact is more common than the symmetrical head-on collision, and oblique frontal impacts must also be taken account of in designing for real-life safety.

From a streamlining viewpoint, we are not now concerned with the interior aspects of occupant protection. What we are interested in are structural developments that will affect the car's exterior dimensions and shape.

Here it is necessary to take a look at the laws of nature that come into play in crash situations. First, there's an equation that's expressed as $KE = 1/2MV^2$. KE means kinetic energy; M means mass and V means velocity. So if you crash a 3,000-pound car at 300 mph, you have to absorb 66,000 foot-pounds.

This enormous force has to be absorbed by the car itself. A barrier does not yield, and even if there are few barrier crashes in real life, the opposing vehicle structure will collapse to absorb its own kinetic energy. Only when a moving car collides with a parked car or another stationary object that offers some degree of compressibility can the energy-absorption task be distributed outside the accident-vehicle.

Now, we have to make a sharp distinction in our minds between energy absorption and occupant protection. The survival of the people inside the colliding car depends only on their personal rate of deceleration during the impact sequence. A human body can survive deceleration forces up to between 40 and 60 g (g is the symbol for the acceleration of a free-falling body in a vacuum, with a standard accepted value of 32.17 feet-per-second-per-second ([ft/sec^2]).

Among the essential conditions for crash survival are survival space. The passenger compartment must be relatively undeformable, and the intrusion of foreign objects, or displacement of the steering column, must be held within strict limits. The driver can be belted in, harnessed, and protected by an automatically inflated air cushion. He will still get killed if the g-loads (proportional to the deceleration rate) exceeds the maximum supportable values.

The Crucial Ride-down Distance

If the car has enough extent of deformable energy-absorbing structure, the deceleration rate can be kept within survivable limits. This structure depends on the bumper system, frame, and sheet metal back to the cowl structure. Maybe there is three feet of collapse in the structure. The crash sequence may be over in 0.3 of a second, but we still talk about it as a sequence, because the various elements collapse in a certain sequence and under a certain load. Naturally, as this occurs, there are peaks in the deceleration rate that exceed the mean value, and these should be leveled off for the smoothest possible ride-down. The engineers call this energy management.

The point I want to make is that there is no substitute for ride-down distance—the distance between the before-and-after length of the car. That's why it is so difficult to build safety into small cars. It's true that small cars are lighter and present a lower energy-absorption problem, but if the ride-down distance is too short, the occupants will be killed from excessive deceleration.

It serves no real purpose to reinforce the vehicle structure and limit body deformation unless the passengers are also assured of not being exposed to lethal g-loads. There must be adequate deceleration distance.

This can easily be provided by adding longer nose sections, but there are practical limits to length. Engineers must learn to work within conventional dimensions, and to "manage" energy absorption by other means—new materials, new support structures, new combinations.

Foam, the Key to Safety Revolution

It is rapidly becoming recognized that the way to go is with foam. Plastic foam is light, and exists in countless varieties, differing in compression rate, elasticity rate, stress-strain curves, and impact-shatter resistance. The Vicom Kinetic Safety Vehicle, first revealed to the public at the New York International Automobile Show in January, 1976, has an astute combination of foam blocks and metallic elements under a plastic body shell.

The Vicom KSV is based on a Ford Pinto, with the entire front end sheet metal, grille and bumper removed. A special bumper is part of the KSV package. But it's the use of the foam

Vicom Kinetic Safety Vehicle (KSV) is based on Ford Pinto. Bumper is fully integrated both front and rear. KSV system assures long deceleration distance with little added weight or overall length.

that's the real secret. The whole thing is built up around a floating bulkhead made of corrugated steel, positioned as a wide-splayed V in front of the radiator. In front and on both sides, there is an interlocking network of semi-rigid steel frames with overlapping sections, linked to the bulk-head. Around these metal elements the foam blocks are disposed in a way that assures maximum concentration of resistance in the direction of the impact forces. They are sequentially arranged so that the foam type used at each specific stage and location provides maximum energy absorption for the impact velocity and strength acting in each section of the total package. But that's not all. The steel units work as guides for the foam blocks, and respond to dislocation under an impact force by pulling in the adjacent foam blocks to make a stronger front against intrusion. The polyurethane molding which serves as the outer skin provides a flexible shell for the containment of the inner structure.

The KSV package absorbs 800 percent more energy than present US "safety" bumpers, and deceleration rates at 40–45 mph impacts are reduced by nearly half. There is no measureable weight gain, and overall length is kept within four to five inches of the original. Body repairs can be made quickly and at low cost.

Created to solve the small-car safety dilemma, the Vicom KSV was designed with improved aerodynamics as a side benefit. The smooth-contoured lengthened nose obviously gives smoother air flow than the stock Pinto. But the point is not what was done—the point is that it has furnished us with proof that advanced energy-management systems can be packaged within compact dimensions and adapted to bodies with improved streamlining without a weight penalty. Once you begin to accept more weight for more passive safety, you begin to compromise the active safety of the car, and that would be a shortsighted, and costly, misplacement of emphasis.

KSV-type packages can also be devised for tail-end impact protection, and steel-and-foam elements may even be used to guard against side-impact penetration. Their use will offer better streamlining opportunities rather than impose restrictions on aerodynamic development.

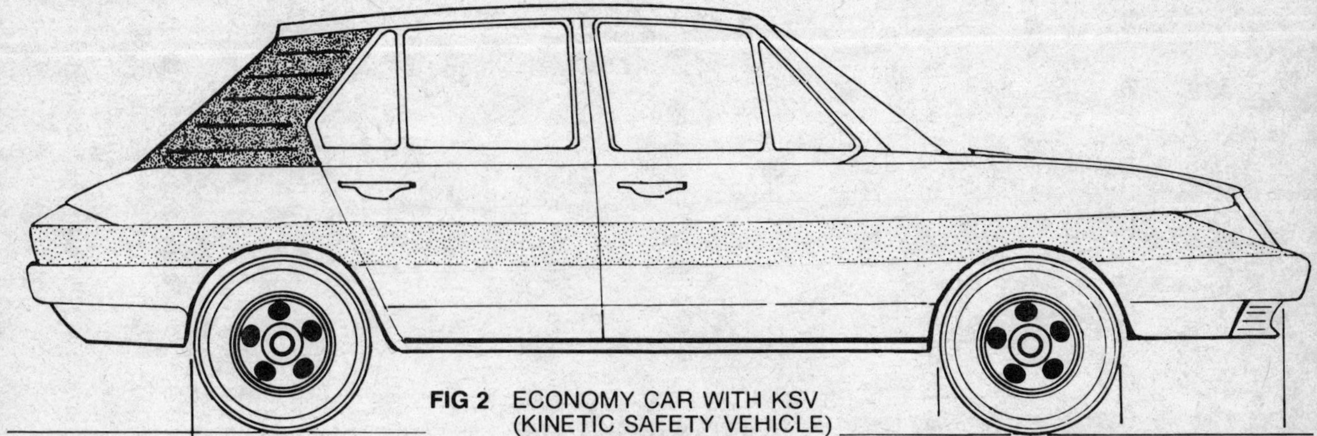

FIG 2 ECONOMY CAR WITH KSV (KINETIC SAFETY VEHICLE) BODY CONVERSION

FIG 1 STOCK SMALL ECONOMY CAR

KSV proposal for Volkswagen Golf show no change in structure of passenger compartment, but added foam packages front, back and sides. For little extra length, streamlining is sensibly improved.

index